The Cell and Molecular Biology of Fish Oogenesis

Monographs in Developmental Biology

Vol. 18

Founded 1969 by *A. Wolsky,* New York, N.Y.

Series Editor
H.W. Sauer, College Station, Tex.

KARGER

Basel · München · Paris · London · New York · New Delhi · Singapore · Tokyo · Sydney

The Cell and Molecular Biology of Fish Oogenesis

Sardul S. Guraya
ICMR Regional Advanced Research Center in Reproductive Biology,
Department of Zoology, College of Basic Sciences and Humanities,
Punjab Agricultural University, Ludhiana, India

68 figures, 1986

KARGER

Basel · München · Paris · London · New York · New Delhi · Singapore · Tokyo · Sydney

Monographs in Developmental Biology

National Library of Medicine, Cataloging in Publication
Guraya, Sardul S., 1930–
The cell and molecular biology of fish oogenesis
Sardul S. Guraya. – Basel; New York: Karger, 1986.
(Monographs in developmental biology, vol. 18)
Bibliography: p
Includes index.
1. Fishes – physiology 2. Oogenesis I. Title II. Series
W1 MO567L v. 18 [QL 639.25 G978c]
ISBN 3-8055-4253-4

Drug Dosage

The authors and the publisher have exerted every effort to ensure that drug selection and dosage set forth in this text are in accord with current recommendations and practice at the time of publication. However, in view of ongoing research, changes in government regulations, and the constant flow of information relating to drug therapy and drug reactions, the reader is urged to check the package insert for each drug for any change in indications and dosage and for added warnings and precautions. This is particularly important when the recommended agent is a new and/or infrequently employed drug.

Printed in Switzerland by Thür AG Offsetdruck, Pratteln
ISBN 3-8055-4253-4

Contents

Dedicated to my wife Surinder

I. Introduction

The fish oogenesis can be divided into various stages such as the oocyte growth including vitellogenesis, ovum maturation and ovulation; ovum maturation and ovulation constitute the terminal stages of oogenesis. Actually, several criteria have been used for staging the process of oogenesis. These include size, amount, and distribution of various cell inclusions, especially yolk granules, and morphology of the chromosomes [*Yamamoto* et al., 1965; *Nagahama,* 1983; *Goetz,* 1983]. The oocyte growth in the fish ovary is a complex process of development and differentiation involving (1) formation of numerous nucleoli, development of lampbrush chromosomes, and inclusion bodies in the nucleus, (2) accumulation of various organelles of diverse morphology, RNA, and inclusions in the ooplasm, and (3) formation of various egg envelopes. All these aspects of oogenesis are characterized by complex morphological, biochemical and molecular transformations that lead to the formation of numerous eggs in the ovary, depending upon the fish species. Many oocytes (or follicles) in different stages of their growth and differentiation are also lost through degeneration (atresia). The intricate nature of developmental changes that occur during fish oocyte growth also led to the expression of very divergent views about the origin, structure, nature and function of some components of the oocyte such as the yolk nuclei, nucleoli (including their extrusion), intranuclear inclusions and lipid bodies, and cortical alveoli in the fish oogenesis; even the modes of yolk formation during vitellogenesis were controversial [*Raven,* 1961]. Keeping in view these controversies as well as economic importance of fish for aquaculture, the past few years have witnessed a great interest in the study of normal development, differentiation, structure, chemistry (including histochemistry), nature and function of various components of growing oocytes in the fish ovary, which have been subjected to new techniques of electron microscopy, histochemistry and biochemistry, including autoradiography and in vitro experiments. Morphological (including ultrastructural) biochemical and physiological (including endocrinological) aspects of vitellogenesis, ovum maturation and ovulation have also been investigated in numerous fish species to develop better methods for induced breeding [*Donaldson and Hunter,* 1983]. The purpose of this monograph is to summarize and integrate the results obtained with such diverse techniques in order to obtain a deeper insight for fish into the cellular, subcellular and molecular biology of oocyte growth and atresia, vitellogenesis, ovum maturation and ovulation in the ovary. My emphasis is thus on the developmental processes of oocyte growth, maturation and ovulation at the cellular and molecular rather than on oogenesis per se [*Tokarz,* 1978] and not much on those environmental events and hormonal mechanisms which regulate various stages of oogenesis. The latter considerations have been addressed in several recent excellent reviews [*Dodd,* 1972; *Reinboth,* 1972; *Donaldson,* 1973; *de Vlaming,* 1974; *Y. Fontaine,* 1975; *M. Fontaine,* 1976; *Lam* et al., 1978; *Richter* et al., 1982; *Fostier* et al., 1983; *Lam,* 1983; *Goetz,* 1983; *Donaldson and Hunter,* 1983; *Liley and Stacey,* 1983], in which very little attention has been paid to the cellular and molecular aspects of oocyte growth and its atresia, vitellogenesis, ovum maturation and ovulation. However, the hormonal factors regulating the synthesis, and transport of exogenous yolk during vitellogenesis as well as regulating ovum maturation and ovulation are discussed in some previous reviews [*Fostier and Jalabert,* 1982; *Ng and Idler,* 1983; *Goetz,* 1983; *Donaldson and Hunter,* 1983].

Integrated and consolidated account of cellular, subcellular and molecular biology of fish oocyte growth and atresia, vitellogenesis, ovum maturation and ovulation, which is not available previously, is essential for a better understanding of the biology of fish egg and its subsequent development, and also to determine the consequences of the influence of various chemicals and radiations at the cellular, subcellular and molecular levels as the fish oocytes at various stages of their growth in the ovary as well as during and after fertilization are being subjected to diverse chemicals due to pollution, and physical agents in the fresh water and sea, possibly affecting fish fecundity and embryology. Unfortunately, reproduction and development in fish occur in an environment increasingly contaminated with many xenobiotic (biologically foreign) compounds. Actually, very little or nothing is known about the mechanisms of action of water pollutants at the subcellular and molecular levels as well as their adverse effects on various components of growing and maturing fish oocyte as well as during embryonic development [*Lam,* 1983] in spite of the fact that in recent years, ecologists and environmentalists have felt the great necessity of the study of effects of chemicals and radiations on fish. *Egami and Ijiri* [1979] have, however, discussed the effects of irradiation on germ cells and embryonic development in teleosts.

A thorough knowledge of biology of fish oogenesis has become extremely important during the recent years because of its close relationship to the development of fish culture. Actually, the normal development, growth and maturation of ovarian oocytes in the fish form the prerequisite for obtaining an adequate supply of fertilized eggs through induced breeding [*Donaldson and Hunter,* 1983], to produce the fry and juveniles for fish culture [*Harvey and Hoar,* 1979], as well as for preservation (including cryopreservation) of ova and embryos [*Stoss,* 1983]; actually cryogenic gene banking of fish gametes has become possible for sperm cells from several species but not for ova and embryos due to the lack of adequate knowledge about their cell and molecular biology.

This monograph will also help us for the better understanding of various aspects of follicular atresia as oocytes at various stages of their growth and differentiation are lost from the ovaries through this degeneration process [*Guraya,* 1976a, 1979a]. The causes of oocyte (or follicle) atresia and its significance are poorly understood not only in fish but also in other vertebrates including mammals [*Guraya,* 1973, 1985]. A thorough understanding of its causes will help us to increase fecundity of the fish by reducing the loss of ova through atresia.

In recent years, the follicular wall (or envelope) of developing oocytes in different fish species has also been extensively studied with various new techniques of histochemistry, biochemistry, autoradiography, electron microscope etc., as its various components such as the zona pellucida (or chorion), follicular epithelium, basal lamina and theca, play some significant roles in the normal growth and differentiation, yolk formation, maturation, ovulation and fertilization of fish eggs, about which our knowledge is still meagre, especially at subcellular and molecular levels; actually pituitary gonadotrophin(s) acts on the follicle wall for stimulating the production of steroid hormones which then regulate vitellogenesis, ovum maturation and ovulation. These studies have demonstrated a considerable diversity in the morphological (or ultrastructural), histochemical and biochemical characteristics of the follicular wall (especially in the zona pellucida) in teleosts having diverse reproductive patterns. The results of different studies, as summarized and integrated here from a comparative point of view, will help us to have a deeper insight into the structure, physiology, maturation, function and interrelationships of various components of the follicular wall in different fish species. An additional purpose of this monograph is also to describe and discuss the cellular sites of steroidogenesis in the follicle wall as well as its gonadotrophic regulation during vitellogenesis, ovum maturation and ovulation in the fish ovary, as the early literature was contradictory [*Guraya,* 1976a, 1978, 1979a]. The ultrastructural, histochemical and biochemical features of steroid-secreting cells in the follicular wall of fish ovary will also be compared and contrasted with those of well-established steroid-gland cell species of mammalian gonads [*Christensen and Gillim,* 1969; *Guraya,* 1971, 1974, 1985; *Christensen,* 1975; *Neaves,* 1975]. Such comparisons will be useful in the better understanding of characteristics and functional significance of cellular sites for steroidogenesis in the maturing follicular wall of oocyte in the fish ovary. The steroid hormones secreted by the follicular wall are now well known to be involved in the (1) synthesis and secretion of exogenous yolk by the liver, (2) ovum maturation and (3) ovulation in the fish [*Wallace and Selman,* 1981; *Fostier* et al., 1983; *Nagahama,* 1983; *Goetz,* 1983; *Ng and Idler,* 1983]. The integrated, fundamental knowledge of cellular and molecular aspects of oocyte growth and maturation in fish, as obtained in this monograph, will form an important basis for encouraging molecular, biological and experimental approaches utilizing state-of-the-art technology, including embryo and genetic engineering. Future advances made through this type of technology are expected not only to maximize the development of fish culture but also to have a deeper insight into developmental processes of fish.

Various components of the ovarian follicle such as nucleus (or germinal vesicle), cytoplasm and its organelles and follicular wall show distinct morphological (including ultrastructural), cytochemical and biochemical changes during previtellogenic and vitellogenic phases of oocyte growth and atresia, ovum maturation and ovulation, which will be discussed in separate chapters here.

II. Nucleus (or Germinal Vesicle)

Since cellular and molecular aspects of oocyte growth are being reviewed here, oogenesis per se or the transformation of oogonium into primary oocytes will not be discussed [see general review by *Tokarz,* 1978]. This transformation has not been well documented in bony fishes, although it is generally accepted that at this time the chromosomes become arrested at diplotene of the first meiotic prophase and the initial follicle is formed. However, the nature of factor(s) involved in meiotic arrest at diplotene needs to be determined. The oocyte at the beginning of its primary growth phase ('chromatin nucleolar stage') consists of a scant cytoplasm and a centrally placed nucleus containing a single, large basophilic nucleolus. The oocyte itself is entirely surrounded by a few squamous follicle cells. Oocytes enter a period of growth which varies from species to species. It is still not known what factor(s) initiate the oocyte growth. These may be hormonal, vascular and nervous in nature. Enlargement of oocytes is caused mainly by the deposition of yolk in their ooplasm as a result of effects of hormones (see Chapter IV).

The oogonium in the fish ovary has a large nucleus and a small amount of cytoplasm. As oocyte growth proceeds, the cytoplasm increases in volume and nucleus becomes enlarged to form the germinal vesicle. Meanwhile multiple nucleoli develop ('perinucleolar stage'), apparently reflecting an amplification of ribosomal genes as will be discussed later on. The morphological (including ultrastructural) changes occurring in the nucleus during previtellogenesis and vitellogenesis are, more or less, similar in different teleosts, suggesting the presence of similar regulatory mechanisms. It is generally accepted now that the nucleus plays a very important role in the regulation of various activities in the cell [*Wischnitzer,* 1973]. Still, our knowledge of its structure as correlated with function in the oocyte is fragmentary and no general conclusions can be drawn from its morphology. Even the significance of polynuclear oocytes observed in the fish ovary [*Kapoor,* 1976a] is not known. The nucleoplasm in the growing oocyte is still difficult to describe in exact terms. Most of the workers describe it as

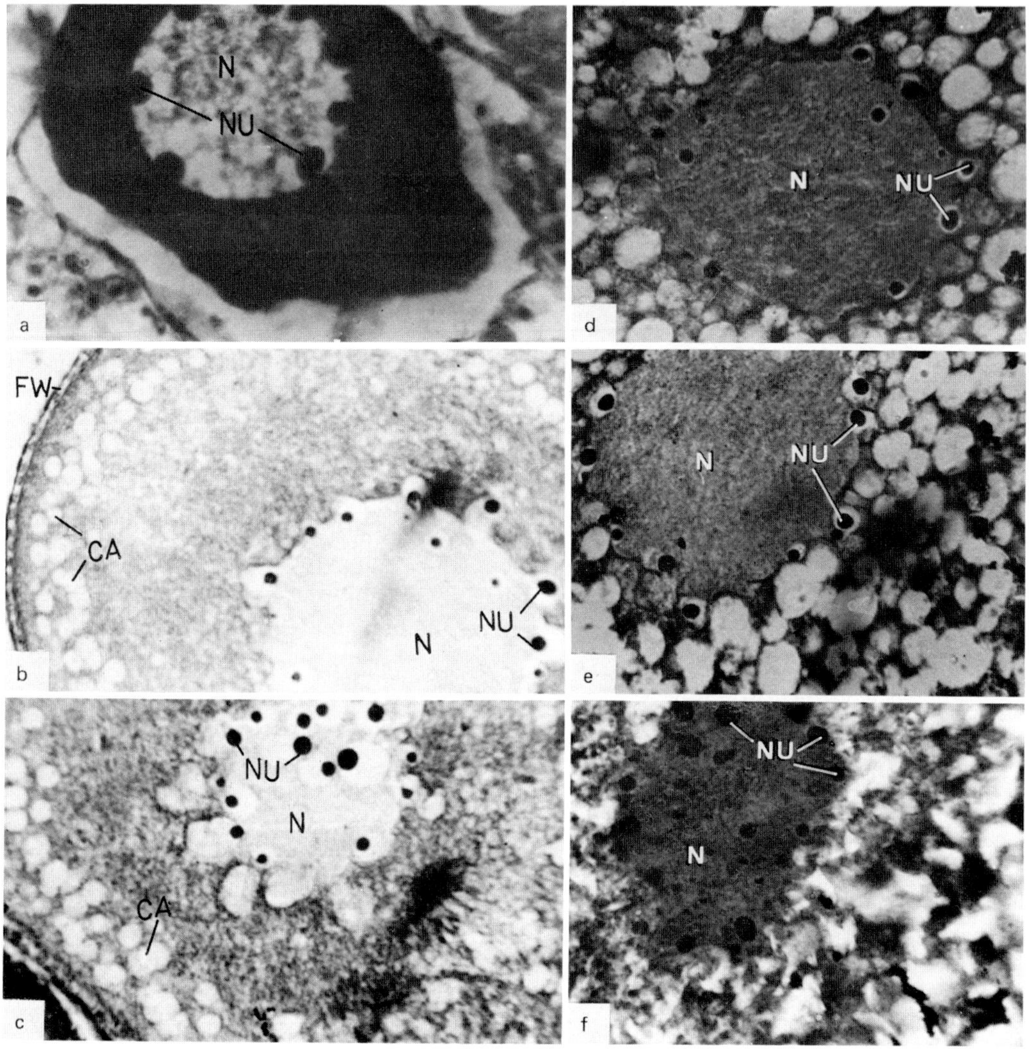

Fig. 1a–f. The development, morphology and distribution of nucleoli (NU) in the nucleus (N) and of cortical alveoli (CA) in the ooplasm of successive stages of oocyte growth in *Colisa fasciatus.* FW = follicular wall. Iron-haematoxylin preparation. *a* × 1,000; *b–f* × 400.

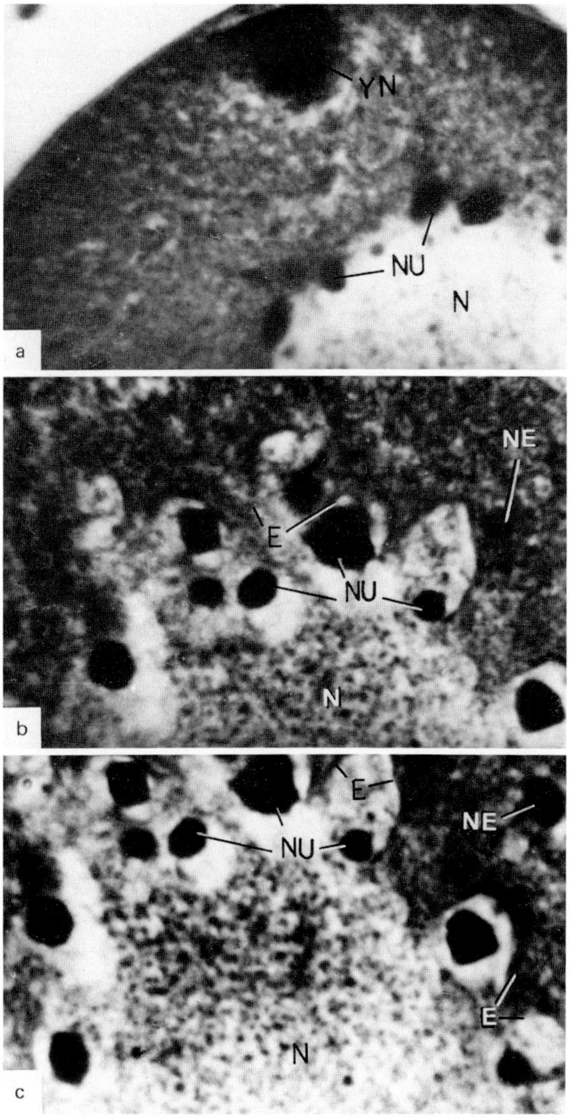

Fig. 2a–c. The morphology and distribution of nucleoli (NU) in the nucleus (N) of early vitellogenic oocytes of *Notopterus notopterus.* Nuclear envelope (E) shows evaginations or pockets. Extruded nucleoli (NE) are seen. Nucleus (N) shows granules of various sizes, which have also accumulated in the perinuclear ooplasm. YN = yolk nucleus. Iron-haematoxylin preparation. × 1,000.

fluffy, granular, or fibrous with multiple nucleoli in the fish oogenesis. Structural, chemical and functional aspects of nucleus in the growing oocyte of fish ovary will be described here in relation to nuclear envelope, chromosomes, nucleoli and their extrusion, intranuclear bodies, etc.

A. Nuclear Envelope

The early differentiation of an oogonium into an oocyte is characterized by the nuclear envelope becoming wavy or undulating (fig. 1a, 2a–c, 3a). With further growth of the oocyte, the nuclear envelope forms large folds and evaginations, resulting in the nucleus becoming lobated (fig. 1b–e, 2b, c); large undulations are usually seen immediately prior to, but also during, vitellogenesis (fig. 1b, c, 2b, c). Obviously, these folds and undulations increase the relative surface of the nucleus for nucleocytoplasmic exchanges of molecules during oocyte growth. Regulation of these morphological alterations of nuclear envelope is not known. Two processes visibly affect the nuclear envelope during the fish oocyte growth. One is concerned with the partial or complete perforation of the envelope, possibly facilitating the passage of nuclear material to the cytoplasm as will be discussed in Section E; the other is the enlargement of the nuclear envelope area as well as its area relative to nuclear volume.

In early oocytes, the nuclear envelope consists of two parallel membranes, separated by the perinuclear space (fig. 3b, 4a–d) [*Droller and Roth,* 1966; *Scharrer and Würzelmann,* 1969a, b; *Upadhyay,* 1977; *Riehl,* 1976a, 1978a; *Azevedo and Coimbra,* 1980]. Some nuclear material, which is often of granular appearance, may adhere to its inner surface (fig. 4a–d). This material originates from 'nucleolar emission' as will be discussed in Section E. Nuclear pores have been described in oocytes of all fish species studied, at least when special attention is paid to the structure of the nuclear envelope (fig. 3–6) [*Droller and Roth,* 1966; *Scharrer and Würzelmann,* 1969a, b;

Fig. 3a, b. Electron micrographs of portion of oocyte nucleus (N) of juvenile rainbow trout. Nucleus is wavy in outline. Nucleoplasm is homogeneous due to uniform distribution of dense granular particles. Nucleoli (NL) are located near the nuclear membrane. Nuclear envelope shows typical nuclear pores (NP). Perinuclear cytoplasm contains mitochondria (M), Golgi apparatus (G), lipid body (LB), and vesicular endoplasmic reticulum. Note the aggregations of dense granules near the nuclear envelope at the perinuclear cytoplasm (arrows) [courtesy of Dr. *S.N. Upadhyay*].

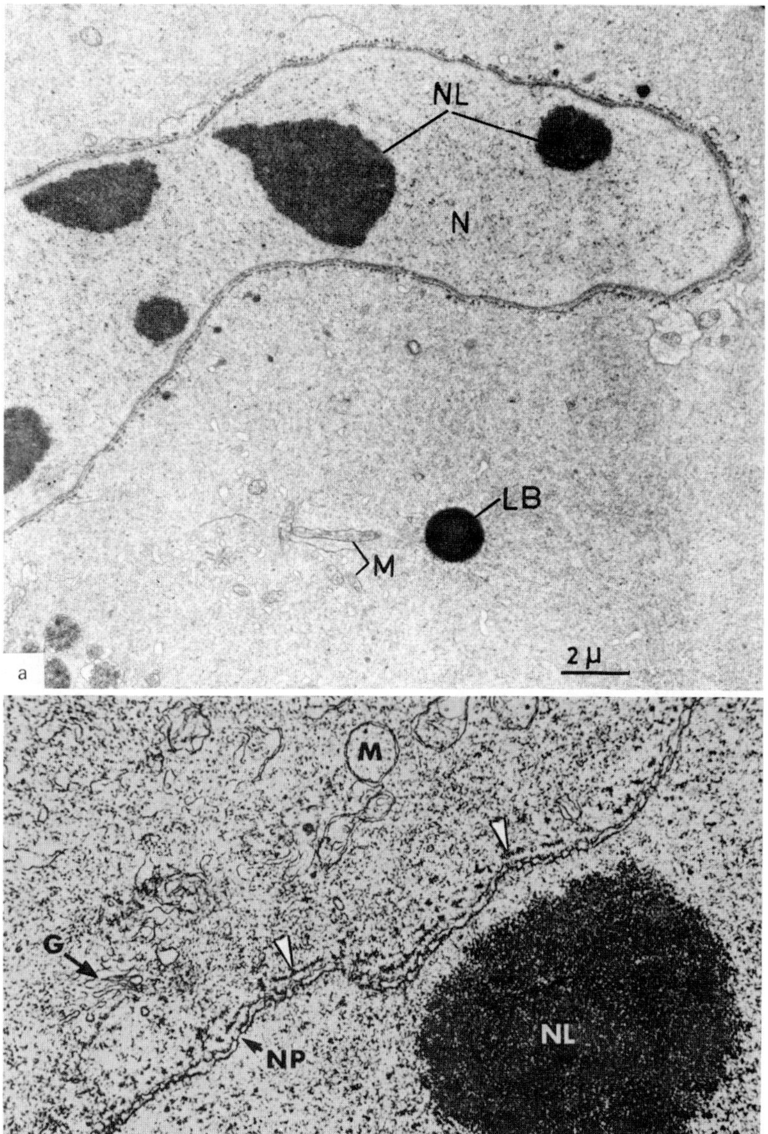
NL
N
LB
M
2 μ
a
M
G
NP
NL
N
1 μ
b

Upadhyay, 1977; *Riehl,* 1976a, 1978a;*Azevedo and Coimbra,* 1980]. Detailed structure of the annuli in the nuclear envelope of oocytes has been studied for a variety of animal species [*Wischnitzer,* 1973]. Most workers have commented on the presence of a central granule in some of the annuli. Central granules are present more often in annuli in early vitellogenetic stages than in mature or nearly mature oocytes. The pore and its associated annular material are known as the pore complex. In an excellent review, *Wischnitzer* [1973] has discussed the development, structure, nature, interrelationships and functions of various components of pore complex in the interphase nucleus of a variety of cell types including oocytes of other animal species.

Cross sections of the nuclear envelope, regardless of the species studied, show that both its membranes fuse in an area approximately 50 nm in diameter (fig. 3b). Whether a diaphragm, formed by the fused membranes, occludes the opening is still a matter of dispute [*Wischnitzer,* 1973]. A close study of tangential sections of the nuclear envelope after osmium fixation has revealed a dense ring, known as the annulus, approximately 20 nm in breadth, around each pore (fig. 6). The shapes and dimensions of nuclear pores and annuli are believed to vary somewhat but the influence of fixation, dehydration, and embedding procedures has to be kept in mind. A detailed study should be made of pore complex at different stages of oocyte growth in fish.

A gradual numerical increase of nuclear pores occurs during the growth of fish oocyte [*Droller and Roth,* 1966; *Scharrer and Würzelmann,* 1969a, b; *Riehl,* 1976a, 1978a; *Upadhyay,* 1977; *Azevedo and Coimbra,* 1980]. These appear in patches which seem to be randomly distributed (fig. 5). The nuclear pores first appear near a mass of mitochondria and intermitochondrial cement that lies close to the nucleus. In older oocytes – and always at the onset of vitellogenesis – nuclear pores cover the entire nuclear envelope so tightly that these appear geometrically arranged mostly in a hexagonal pattern [*Droller and Roth,* 1966; *Scharrer and Würzelmann,* 1969a, b; *Upadhyay,* 1977; *Azevedo and Coimbra,* 1980]. No oocyte beyond vitellogenesis is described as being without nuclear pores.

The nuclear membranes may form 'blebs' (fig. 4a, 7a, b) which are pinched off so as to become detached from the nuclear envelope [*Scharrer and Würzelmann,* 1969a, b]. These blebs finally form part of the endoplasmic reticulum. Blebbing from the inner nuclear membrane into the nucleoplasm has also been reported [*Azevedo and Coimbra,* 1980]. Ribosomes are not attached to these blebs. Blebbing of the outer nuclear membrane has

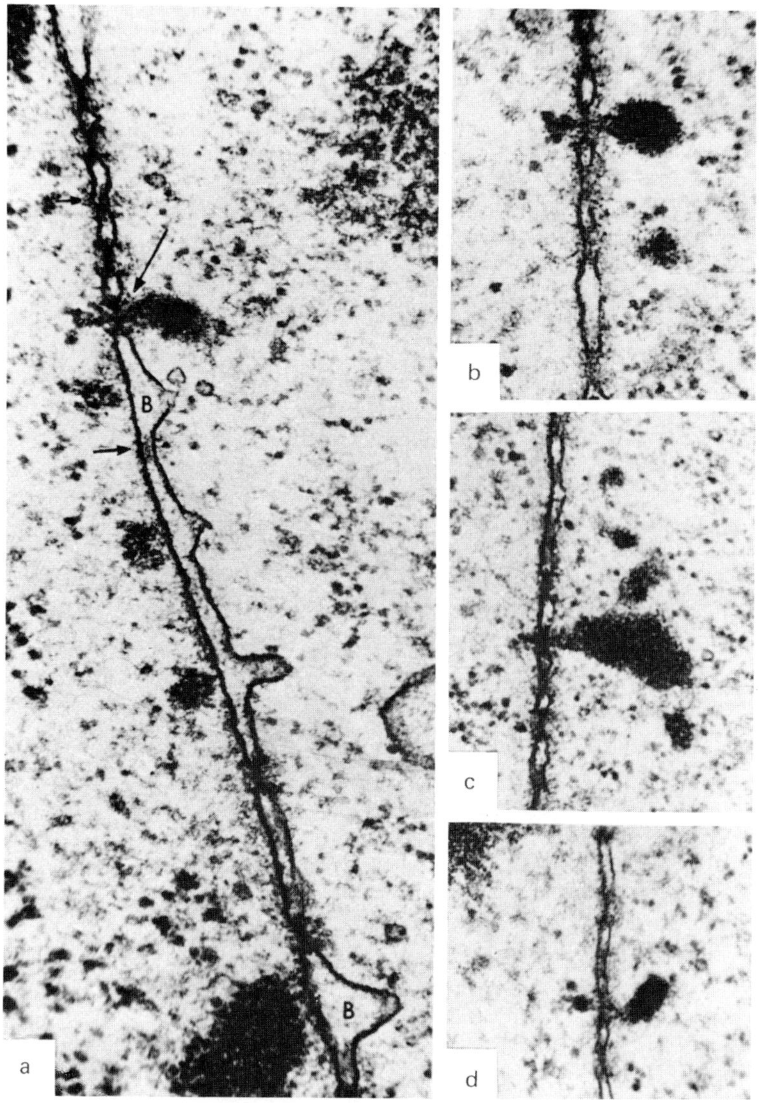

Fig. 4. Electron micrographs illustrating nucleolo-cytoplasmic transport in the oocyte of lungfish. Nucleus on left, cytoplasm on right of nuclear envelope. *a* Numerous small aggregates of nucleolar material close to the nuclear boundary. One of these (long arrow) continues into the cytoplasm via one of several nuclear pores (shown in transverse section) in the area. Note also blebs (B) projecting toward the cytoplasm, and presumed former sites of nuclear pores (short arrows). *b–d* Other examples of cytoplasmic accumulations of presumed nucleolar material in continuity with small intranuclear aggregates via nuclear pore complexes. *a* × 63,000; *b* × 68,000; *c, d* × 50,000 [from *Scharrer and Würzelmann,* 1969a].

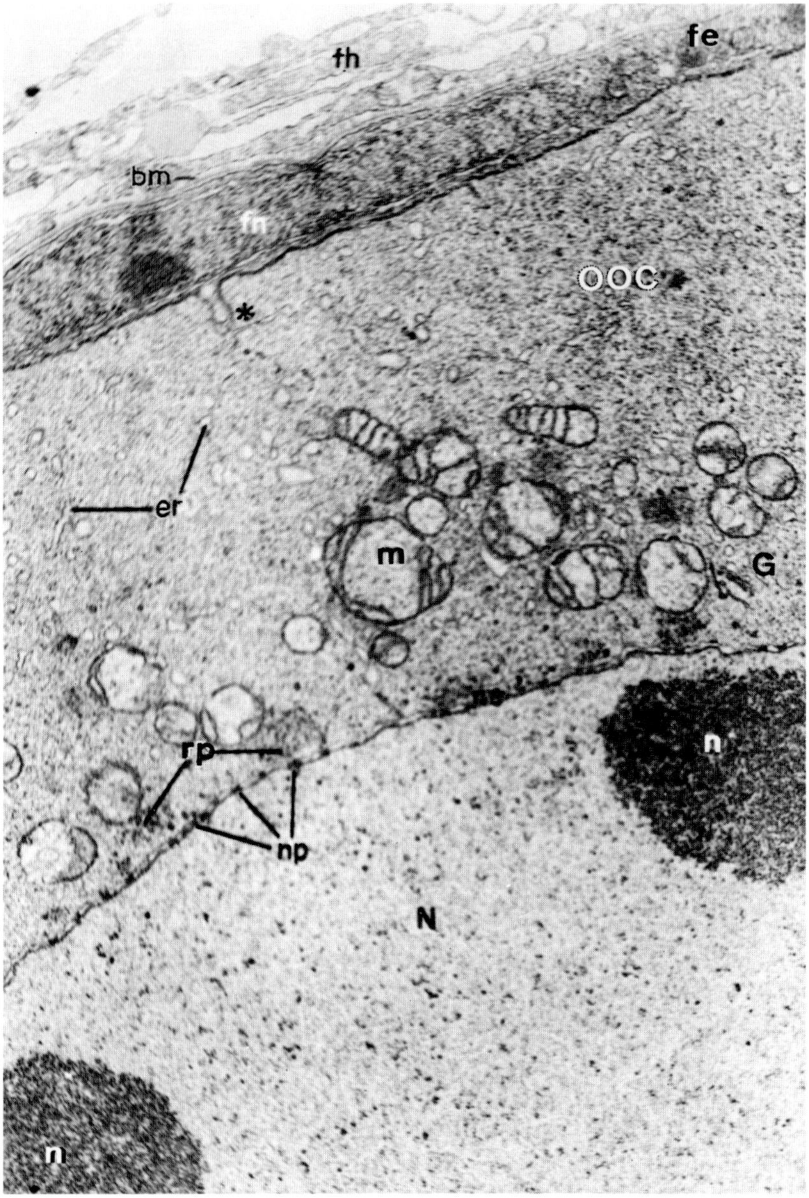

Fig. 5. Developing guppy oocyte (OOC) surrounded by a simple squamous follicular epithelium (fe) with flattened nuclei (fn), a basement membrane (bm), and a sheath of connective tissue (th). Follicle cell process (*) is seen. Oocyte shows endoplasmic reticulum

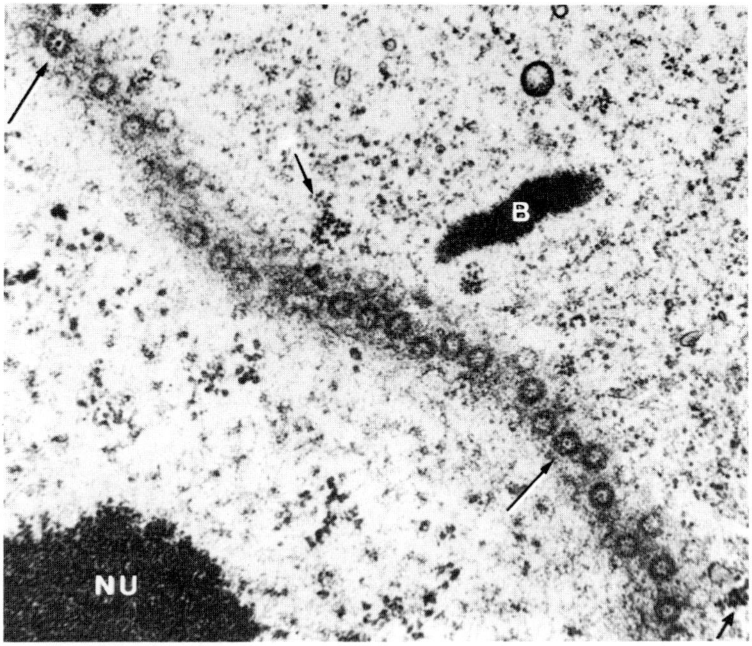

Fig. 6. Nuclear envelope of lungfish oocyte in tangential section, shows closely spaced annuli and dense 'dots' in centers of nuclear pores (long arrows). Aside from dispersed electron-dense particles, the cytoplasm contains an elongate body (B) presumably derived from nucleolus (NU) on opposite side of nuclear boundary. Also note clusters of ribosome-like particles in perinuclear ooplasm (short arrows). × 34,000 [from *Scharrer and Würzelmann,* 1969a].

(er) and many free ribosomes; Golgi bodies (G) and large mitochondria (m) are localized in the perinuclear ooplasm. At the outer surface of the nuclear envelope (ne), especially at the nuclear pores (np), there appear dense granular areas (rp) which are similar in texture to the large nucleoli (n) at the periphery of the spherical nucleus (N). × 18,000 [from *Droller and Roth,* 1966].

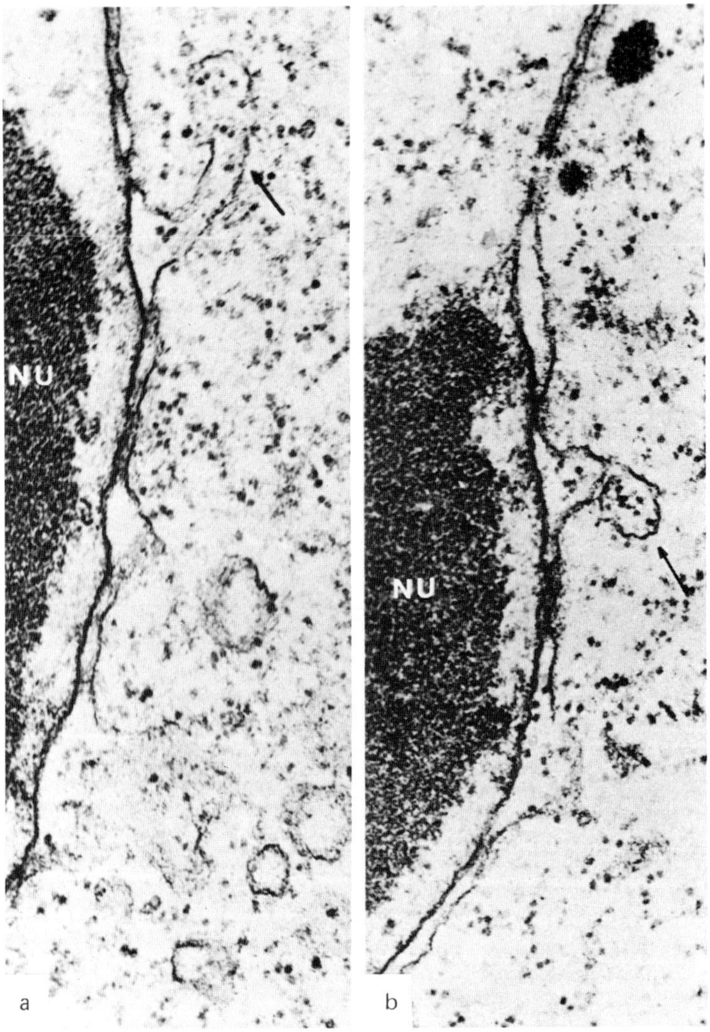

Fig. 7. a, b Examples of blebbing of nuclear envelope next to nucleolar bodies (NU) in the oocytes of lungfish. Note especially elongated projections (arrows) of perinuclear cistern, containing electrondense particles presumed to be in transit to the ooplasm. Adjacent intracytoplasmic-dense particles seem to have spilled out from the intracisternal compartment through openings in the outer nuclear membrane. × 50,000 [from *Scharrer and Würzelmann,* 1969a).

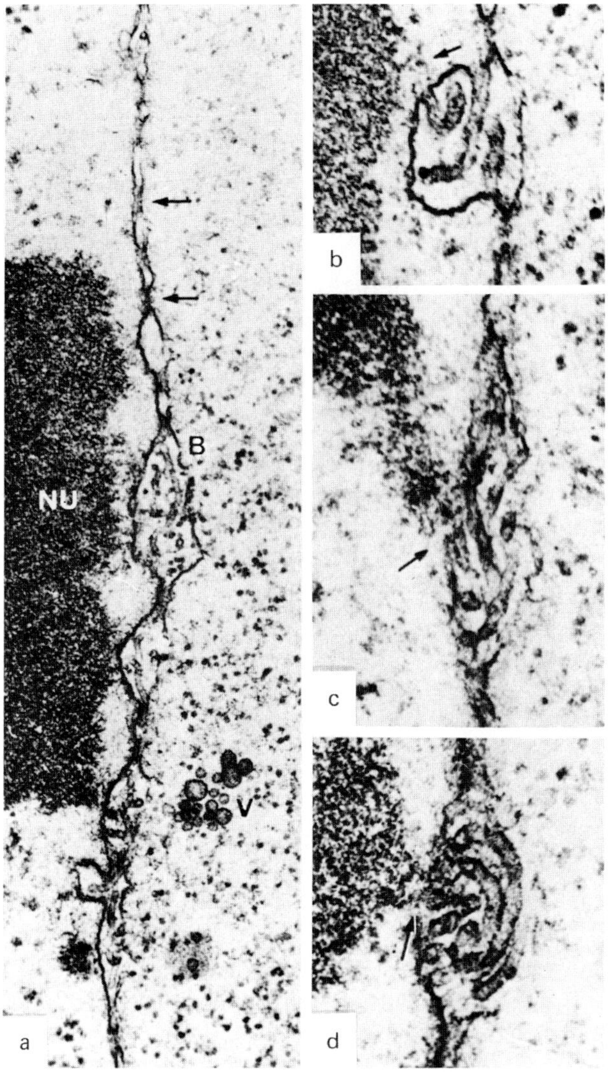

Fig. 8. Nuclear envelope from the oocytes of lungfish. *a* Part of nuclear envelope close to nucleolar body (NU) showing blebs (B) containing microtubules and small particles in close association with areas of contact between nucleolus and inner nuclear membrane. Note vesicles (V) and electron-dense particles in perinuclear cytoplasm, and nuclear pores (arrows). *b–d* Examples of blebs in perinuclear cistern suggesting communication between microtubular content and nucleolar material (arrows). Note partial absence of outer nuclear membrane (cytoplasm side at right). *a* × 40,500; *b, c* × 68,000; *d* × 88,000 [from *Scharrer and Würzelmann,* 1969b].

been, more or less, convincingly reported [*Scharrer and Würzelmann,* 1969a, b; *Riehl,* 1976a, 1978a]. Usually ribosomes are attached to the cytoplasmic blebs. The outer nuclear membrane is devoid of ribosomes, while the blebs belong to the rough type. The endoplasmic reticulum vesicles lying nearest to the nuclear envelope are devoid of ribosomes while ribosomes are present on the more distant vesicles. The blebs are finally included in the endoplasmic reticulum of the growing oocyte. Regulation of bleb(s) formation is not known.

The blebs of the perinuclear cistern, for a limited period during oogenesis of African lungfish *(Protopterus aethiopicus),* contain a special kind of microtubular elements in addition to other inclusions (fig. 8a–d). Most of these blebs face parts of multiple nucleolar bodies that extend toward and make contact with the inner nuclear membrane. The microtubular lumen contains a finely dispersed material of moderate electron density which appears to be in contact with the nucleolar material (fig. 8d). Some microtubular arrays without apparent connection with the nuclear envelope are also seen within both the perinuclear cytoplasm and nucleoplasm. Intracisternal microtubular apparatus is believed to be related to transport of special nucleolar components to the cytoplasm and to break down of parts of the nuclear envelope to its restitution during the subsequent stages of oogenesis.

B. Chromosomes

With the initiation of oocyte growth in the fish ovary, the chromosomes undergo conspicuous morphological, histochemical and biochemical changes. The nucleus of oogonia and very young oocytes shows chromosomes (chromatin substance) which stain for DNA and proteins [*Kraft and Peters,* 1963]. With the oocyte growth, the positive histochemical reaction of chromatin for DNA becomes negative [*Rastogi,* 1968a]. Meanwhile, the chromosomes develop a lampbrush appearance and thereafter are called

---

Fig. 9. From the oocytes of *Brachydanio. a, b* Lampbrush chromosomes in living oocyte nuclei studied with phase contrast. The loops of the lampbrush chromosomes are collapsed by fixation. *a* Stage 1; nucleus with lampbrush chromosomes; nucleoli appear as dark granules. *b* Stage 2; part of a nucleus; arrow: giant loop axis of a lampbrush chromosome. *c–g* ^{3}H-cytidine-^{3}H-uridine incorporation into oocyte nuclei of different stages studied with phase contrast. *c* Oocyte of late stage uniformly labelled nucleus. Labelling of

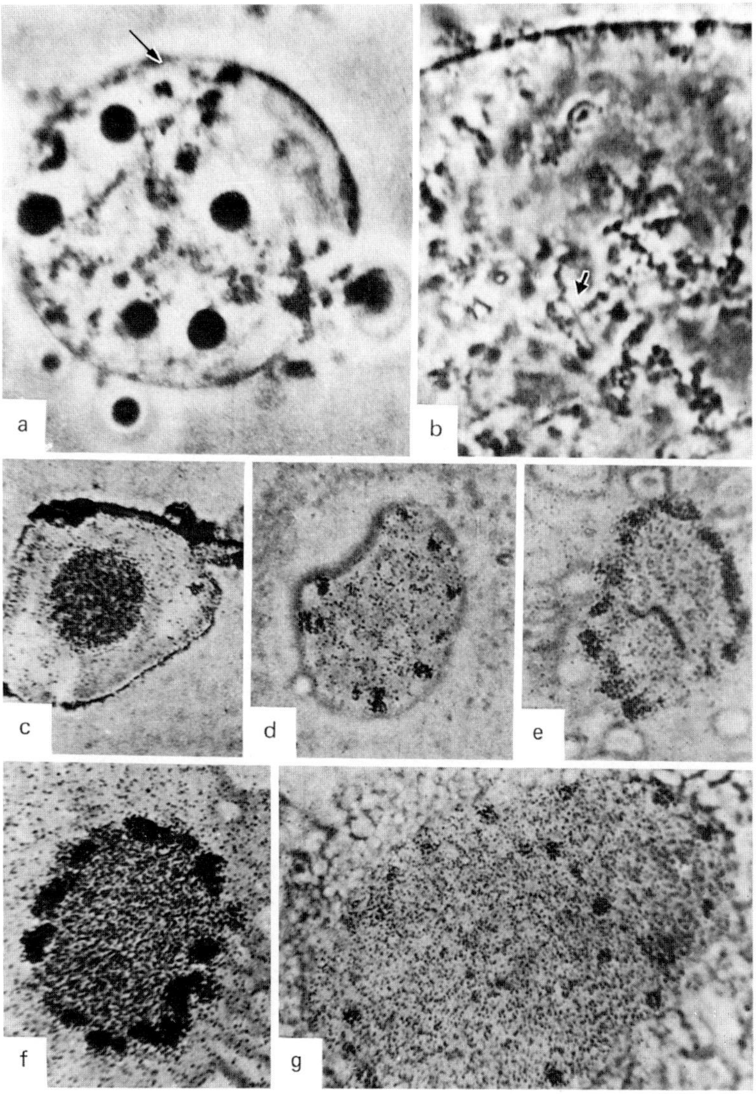

9

nucleoli is not particularly pronounced. *d* Oocyte nucleus of late stage 2. Besides the labelled nucleoli, there is a distinct label over the interior parts of the nucleus. *e* Oocyte nucleus of stage 3. *f* Oocyte nucleus of stage 3. A further prolongation of the incorporation period has no effect on the labelling pattern. *g* Oocyte nucleus of stage 4. In this stage only little labelling can be achieved even by a long-time incubation. *a* × 750; *b* × 1,200; *c–e* × 250; *f* × 350; *g* × 250 [from *Baumeister,* 1973].

lampbrush chromosomes. Negative results of the Feulgen and pseudoisocyanin reactions for DNA may be explained by the fact that chromosomes in the lampbrush stage are largely despiralized [*Hess and Meyer,* 1963]. The lampbrush chromosomes observed during the diplotene stage in different teleosts [*Braeckevelt and McMillan,* 1967; *Lehri,* 1968; *Baumeister,* 1973; *Raikova,* 1976; *Wourms,* 1976; *Ramadan* et al., 1978; *Monaco* et al., 1980], are indicative of intensive RNA synthesis in the nucleus of growing oocyte [*Davidson,* 1976].

Because of their extended configuration, lampbrush chromosomes are not readily visible in routine histological preparations, however, *Bara* [1960] showed their presence in the oocytes of the mackerel *(Scomber scomber)* and their disappearance immediately prior to germinal vesicle breakdown during maturation. *Baumeister* [1973] reported more clearly the presence of lampbrush chromosomes in teleost (zebra fish *Brachydanio rerio*) (fig. 9a, b). In its young oocytes, the lampbrush chromosomes show a very fine structure so that the cytochemical demonstration for DNA becomes impossible. Actually, these are small and relatively frail (fig. 9a, b). Since chromosome length and loop circumference depends on species specific levels of DNA [*Ullerich,* 1970] one may expect correspondingly low values in *Brachydanio.* The values determined for teleostei range between 0.7 and 1.9×10^{-12} g DNA/haploid set [*Hinegardner,* 1968; *Rasch* et al., 1970]. The Amphibia have on the average 20- to 100-fold higher amounts of DNA [*Ullerich,* 1970], compared with teleosts. The activities of DNA polymerases, α and β, have been demonstrated in the extracts of germinal vesicle (nuclei) isolated from the maturing oocytes of the loach, *Misgurnus fossilis* [*Mikhailov* et al., 1983]. Nearly 80% of the total α-polymerase activity is associated with germinal vesicle while the γ-polymerase activity is revealed in the cytoplasm. The nonhistone nuclear proteins accumulate in the germinal vesicle during oocyte maturation as revealed with autoradiography. Further extensive studies must be made of DNA and RNA polymerases, other enzymes and proteins in the germinal vesicle during different stages of fish oogenesis. The results of such studies will be very useful for the better understanding of regulatory mechanisms involved in transcription.

One of the most favourable materials for the study of the origin of RNA from DNA is seen in classic oocyte lampbrush chromosomes, especially of amphibians [*Callan,* 1963; *Davidson,* 1976]. *Duryee* [1950] first observed that a marked increase in the volume of oocyte cytoplasm and nucleus occurs when the chromosomes are in their extended lampbrush form as also extended and confirmed by *Callan and Lloyd* [1960]. This

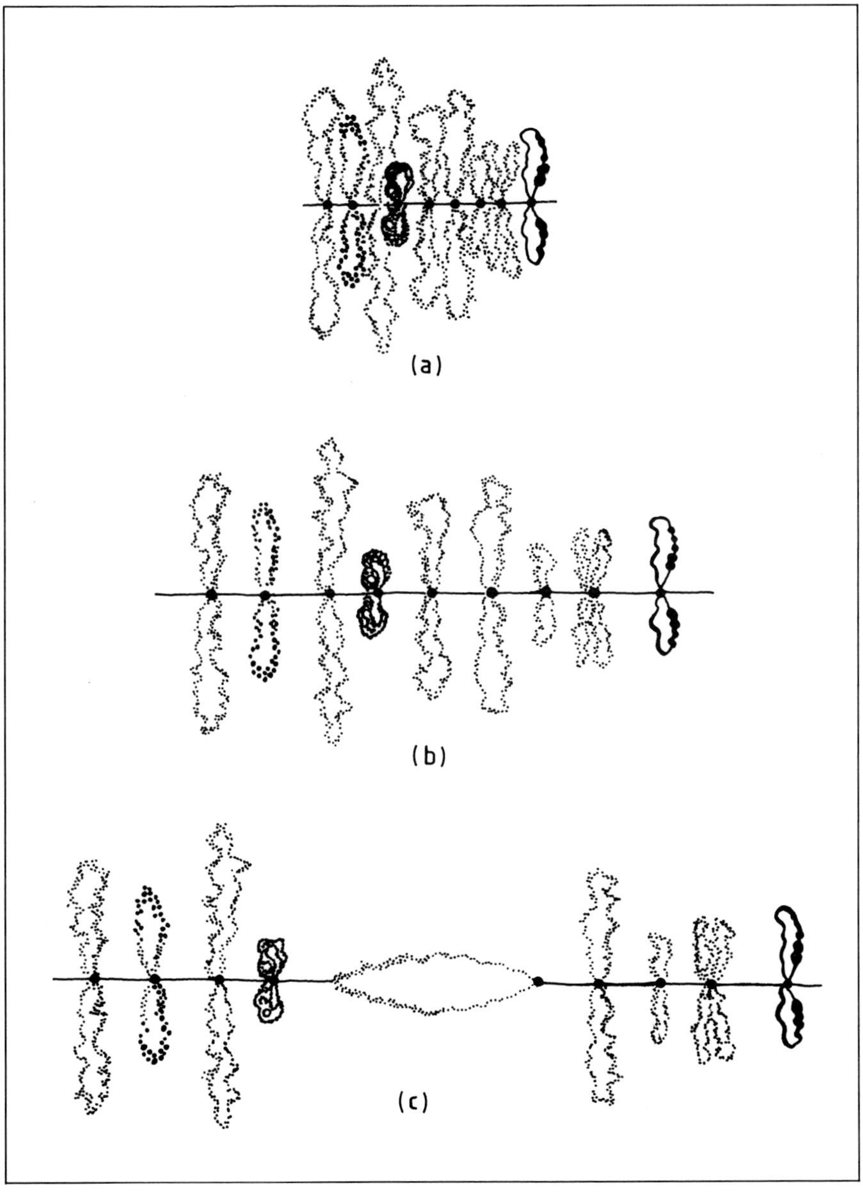

Fig. 10. Diagrams to illustrate what happens when part of a lampbrush chromosome is stretched: (*a*) unstretched; (*b*) stretched within the elastic limit; (*c*) stretched beyond the elastic limit – one chromomere has broken, and a pair of lateral loops span the break [from *Callan,* 1963].

suggested that the morphological changes in these chromosomes during oocyte growth are probably indicative of the synthesizing activity of the genes. Each chromosome is composed of two homologs of enormous length (fig. 10). The axis of each consists of a thin, elastic chromonema along which are located many chromomeres. Extending from the chromosomes are lateral loops which show characteristic morphology (fig. 10) [*Callan,* 1963; *Davidson,* 1976]. The chromomeres and lateral loops contain DNA. In addition, the matrix surrounding the lateral loops consists of RNA and protein. Various autoradiographic studies have shown that these two substances are actively synthesized at the loops and eventually become detached to lie free in the nucleoplasm [*Davidson,* 1976].

The presence of lampbrush chromosomes in the nucleus of fish oocyte also suggests that they perform the same basic function of production of protein and RNA. The high rate of RNA synthesis, which is characteristic of the lampbrush chromosomes of amphibian oocytes [*Davidson,* 1976], is also observed in the previtellogenic oocytes of *Brachydanio* (fig. 9c–h). The uniform labelling of the nucleoplasm in nuclei of early oocytes may be another indication of the extended structure of the chromosomal material. At the beginning of exogenous yolk accumulation at stage 4, the nonamplified part of the oocyte genome becomes inactive [*Korfsmeier,* 1966]. The lampbrush chromosomes in acipenserid oocytes also produce both proteinaceous spheres and ribonucleoprotein bodies [*Raikova,* 1976]. The synthesis of RNA on lampbrush chromosomes is a DNA-dependent process, as determined by its sensitivity to actinomycin D, which also inhibits RNA synthesis in the nucleolus. Actinomycin D apparently 'blocks RNA synthesis by selectively combining with the "primer" DNAs needed for the RNA polymerase reaction' [*Davidson,* 1976]. From the results of various studies as discussed here, it can be concluded that the nucleus of growing fish oocyte develops lampbrush chromosomes which form important sites for the synthesis of RNA and protein (fig. 9a–h). The molecular mechanisms which regulate the formation of lampbrush chromosomes are still to be determined. Recombinant DNA technology, that includes a series of powerful approaches for elucidating the structure, organization, and regulation of genetic sequences [*Stein and Stein,* 1984], must be effectively utilized to determine the regulation of some specified genes in the lampbrush chromosomes of fish oogenesis. A thorough study of nuclear (or chromosomal) proteins during oogenesis will help us in the elucidation of regulation of stage-specific gene activity (or transcription), as our knowledge is very meagre in this regard.

C. Nucleoli

1. Morphology (Including Ultrastructure), Cytochemistry and Biochemistry

Initially, one or two nucleoli are present in the nucleus of young oocytes (fig. 11a). As the oocyte grows and differentiates further, there is seen a corresponding rapid increase in the number of nucleoli (fig. 1a, 2a, 3a, 11b–f). Meanwhile, these start lying inner to the nuclear envelope where they finally form a regular layer. Multiple nucleoli apparently reflect an amplification of genes [*Vincent* et al., 1969; *Vlad,* 1976; *Monaco* et al., 1980], which must be established with recombinant DNA techniques [*Stein and Stein,* 1984]. The regulation of amplification of genes needs to be determined at the molecular level.

The phases of meiotic prophase during fish oogenesis are generally distinguished by the location of the prominent nucleolus. The first growth phase of oocytes, corresponding to meiotic prophase, is usually divided into three stages: (1) the chromatin nucleolus stage; (2) the early perinucleolus stage, and (3) the late perinucleolus stage [*Guraya* et al., 1975, 1977; *Tokarz,* 1978]. The chromatin nucleolus stage can be distinguished by a conspicuous nucleolus connected with chromatin threads. In this stage, presynaptic oogonia undergo synapsis (during the zygotene stage) and form postsynaptic oocytes. In the early perinucleolus stage, the size of the nucleus is increased as the oocytes, and the nucleoli increase in number and size and are placed at the periphery of the nucleus. Meanwhile, Balbiani's vitelline body (or the so-called yolk nucleus) is developed as will be discussed in Chapter III. The late perinucleolus stage (or diplotene stage) can be distinguished from the previous stage mainly by the increasing size of the oocytes. Lampbrush chromosomes developed during the diplotene stage are already discussed in Section B.

In number, size, shape and structure, the nucleoli vary greatly in different teleosts and may lie randomly within the germinal vesicle (fig. 1a–f, 2a–c, 3a, 11a–f). *Sobhana and Nair* [1977] have reported the multiplication and fate of nucleoli in the oogenesis of olive carp *(Puntius sarana).* An electron microscope study of *Droller and Roth* [1966] revealed that each nucleus at first in the guppy *(Lebistes reticulatus)* contains a single nucleolus and in the later stages of oocyte growth it develops several prominent nucleoli (fig. 5). *Scharrer and Würzelmann* [1969a] also observed an impressive development of the nucleolar apparatus in growing oocytes of the African lungfish (fig. 12). Numerous nucleolus-like bodies establish close

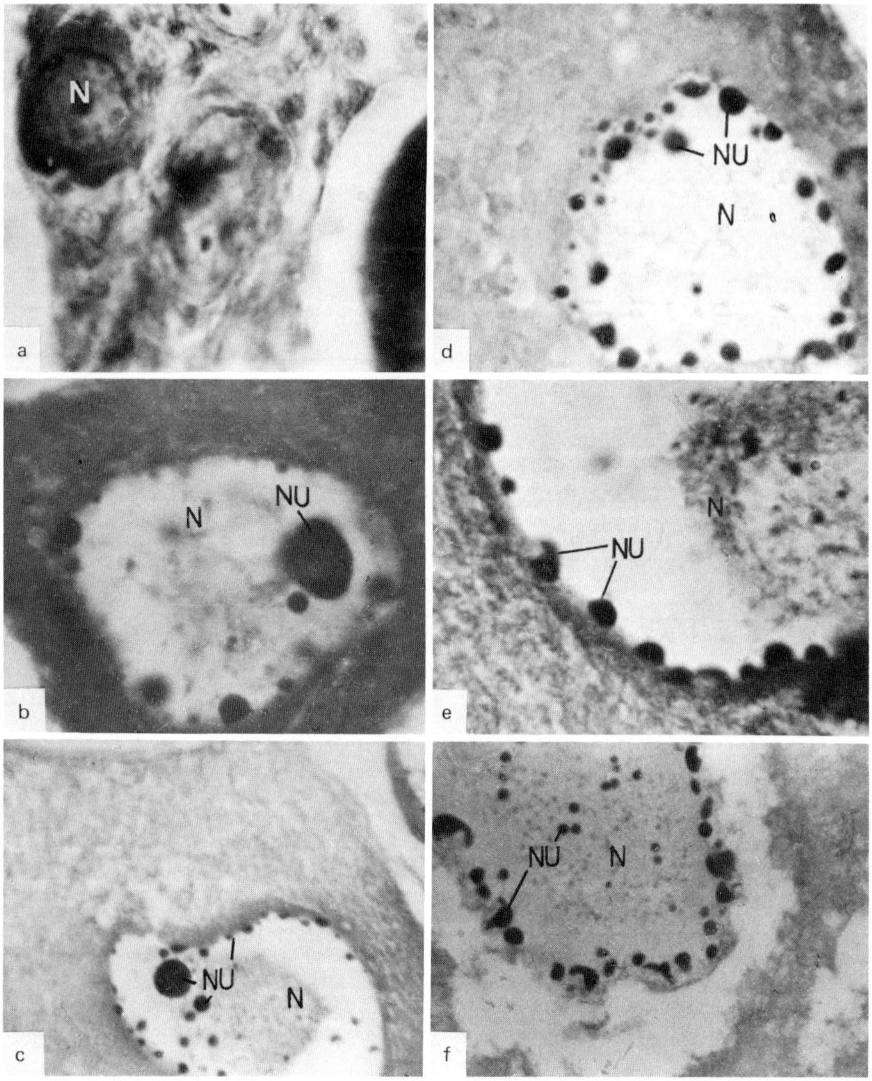

Fig. 11. a–f The development, morphology and distribution of nucleoli (NU) in the nucleus (N) of successive stages of oocyte growth in *Cyprinus carpio.* Eosin-haematoxylin preparation. *a, b* × 1,000; *c–f* × 400.

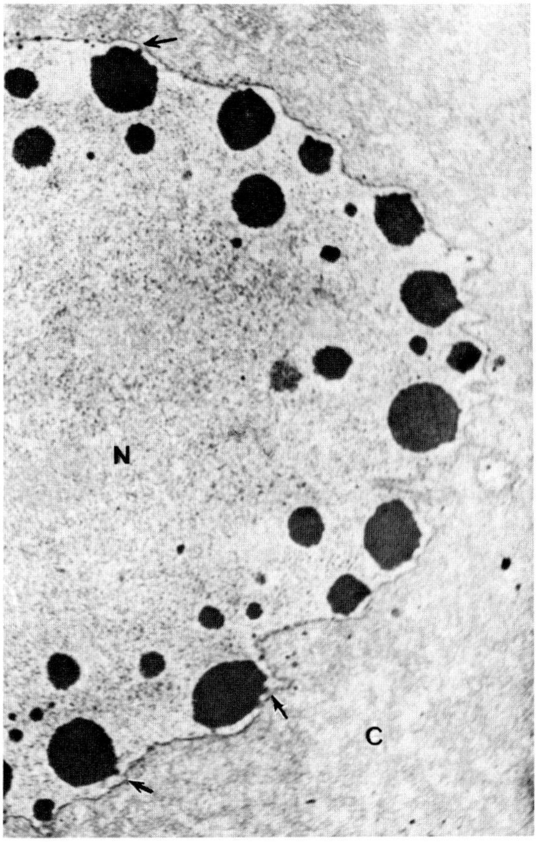

Fig. 12. Low-power view of young oocyte of lungfish showing alignment of numerous compact nucleolar bodies close to undulating nuclear envelope. N = nucleus; C = perinuclear cytoplasm. Small pedicels are indicated by arrows. × 7,400 [from *Scharrer and Würzelmann,* 1969a].

spatial relationships with the nuclear envelope by extending pedicels and streams of finely dispersed material towards its inner membrane (fig. 13). *Baumeister* [1976] using light and electron microscopy and autoradiography studied the formation of multiple nucleoli in growing oocytes of the zebra fish *(Brachydanio rerio).* In oogonia only one nucleolus is observed. Gene amplification occurs during stages 1–3 of 5 developmental stages until 1,000 nucleoli/oocyte nucleus develop. Multiple nulceoli separate from the chromosomes and migrate as compact bodies to the nuclear envelope where their structure becomes loose. *Azevedo and Coimbra* [1980] have recently studied the evolution of nucleoli in the oogenesis of a viviparous teleost *(Xiphophorus helleri)* with electron microscopy and autoradiography. Its germ cells contain one or two nucleoli which do not multiply but grow in size considerably and meanwhile undergo marked vacuolization. The presence of nucleoli throughout oocyte growth is ubiquitous.

In earlier stages of oocyte growth, the nucleoli generally stain homogeneously in different stains (fig. 1–3, 11, 12), but later on they develop a vacuolated appearance due to the presence of one to several vacuoles of variable size (fig. 13–15) [*Scharrer and Würzelmann,* 1969a; *Azevedo and Coimbra,* 1980]. The vacuoles either do not stain or show very weak staining. The nucleoli of vitellogenic oocytes stained with iron haematoxylin show a very heterogeneous structure due to the development of vacuoles and dense granules of variable size [*S.S. Guraya,* unpubl. observations]; the granular components stain more intensely with iron haematoxylin than the rest of the nucleolar substance.

Corresponding to the multiplication and growth of nucleoli, there occurs an extensive development and accumulation of granular substance throughout the nucleoplasm (fig. 1d–f, 2a–c, 16a–e). In its staining reactions, the granular substance resembles closely the nucleoli, showing some close relationship between them. Various electron microscope studies have also revealed that the nucleolus is an accumulation of granules, similar to those seen in the nucleoplasm (fig. 17, 18). The size of these granules usually ranges between 10 and 20 nm, most often 15 nm. The nucleolus generally consists of a dense mass of ribosome-like particles 15–20 nm in diameter, intermixed with fine 'granular-fibrillar' material [*Droller and Roth,* 1966;

Fig. 13. a Nucleolar body with central cavity from the oocyte of lungfish. Pedicels and streams of dispersed material seek contact with inner nuclear membrane. At these points (arrows) the perinuclear cistern shows widened gaps, some of which seem to give rise to cytoplasmic vesicles (V). Note affiliation between such vesicles and mitochondria (M). *b, c* Nucleoli sending delicate streams (long arrows) to nuclear envelope and elongate small bodies (B) of the same appearance on the cytoplasmic side in the oocyte of lungfish. *b* Note origin of cytoplasmic vesicles from outer nuclear membrane (short arrow). *c* Note mitochondrial elements (M), small vesicles (V), and nuclear pore (short arrow). In area indicated by A, material of medium density spreads out within perinuclear cistern. *a* × 32,000; *b* × 47,000; *c* × 34,000 [from *Scharrer and Würzelmann,* 1969a].

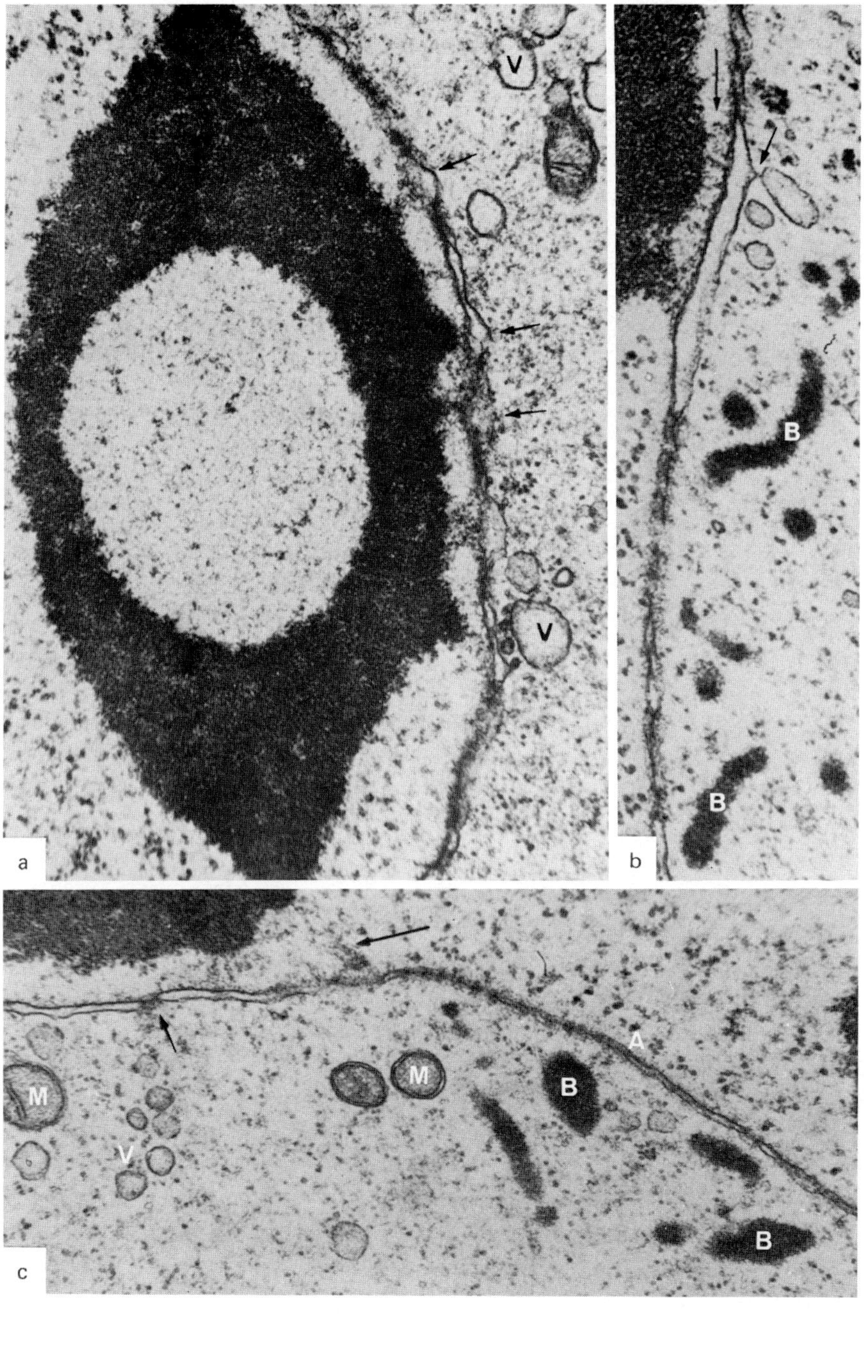

13

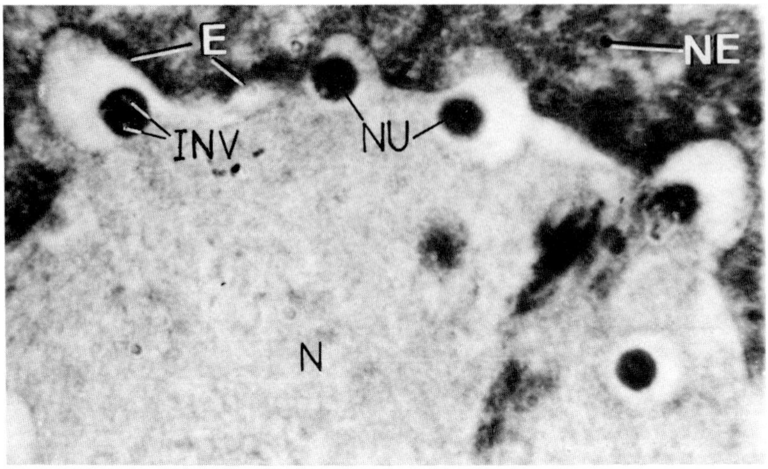

Fig. 14. Vacuoles (INV) in the nucleoli (NU) of nucleus of early vitellogenic oocyte in *Colisa fasciatus.* Nuclear envelope (E) forms evaginations or pouches. Small extruded nucleolus (NE) is also seen. Iron-haematoxylin preparation. × 1,000.

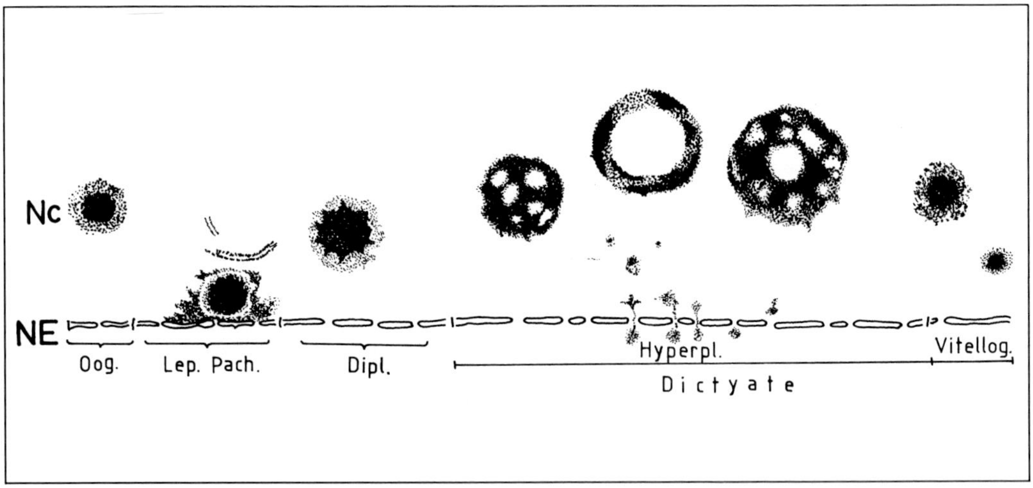

Fig. 15. Evolution of nucleolus (Nc) during oogenesis in *Xiphophorus.* NE = nuclear envelope [from *Azevedo and Coimbra,* 1980].

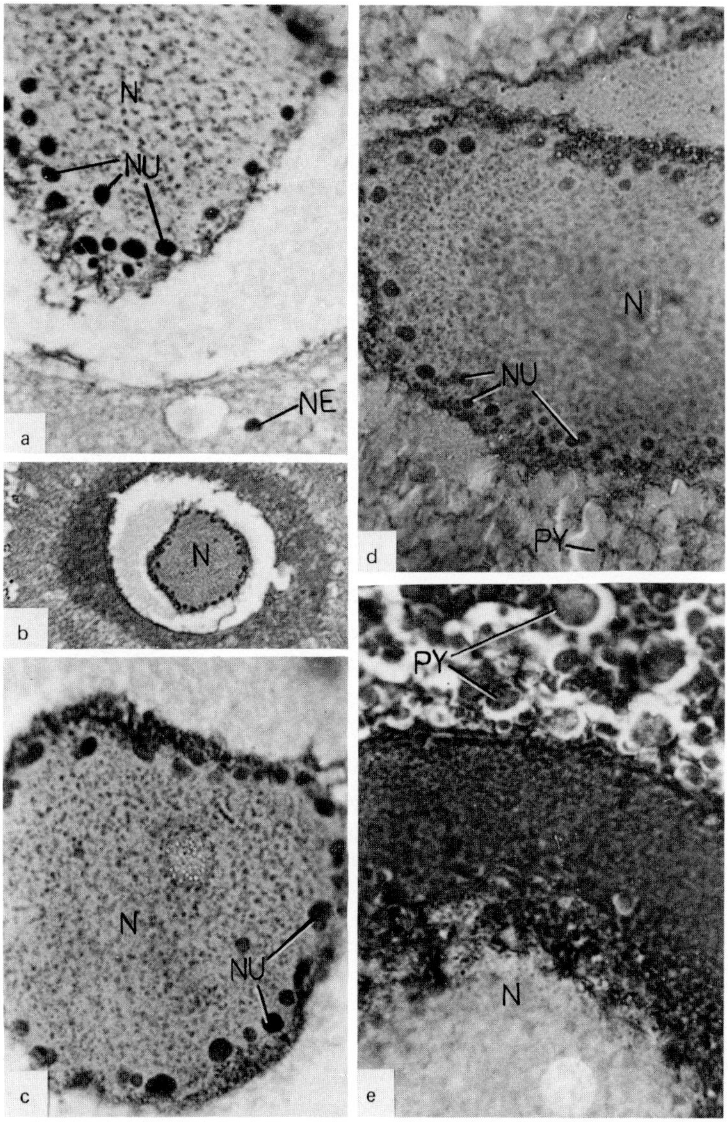

Fig. 16a–e. The morphology and distribution of nucleoli (NU) and accumulation of granules of various sizes in the nucleus (N) of the successive stages of vitellogenic oocytes in *Cyprinus carpio.* Accumulations of granules similar to those of nucleus are also seen in the perinuclear cytoplasm as shown in *b. a* extruded nucleolus (NE). *d, e* Proteid yolk platelets (PY). Eosin-haematoxylin preparation. *a* × 300; *b* × 75; *c, d* × 300; *e* × 750.

Scharrer and Würzelmann, 1969a, b; *Azevedo and Coimbra,* 1980]. Vacuoles in the nucleolus seem to consist primarily of this 'granular-fibrillar' material. Oogonia and oocytes of *X. helleri* have compact nucleoli up to the end of diplotene with a central fibrillar core and granular periphery (fig. 19a, b) [*Azevedo and Coimbra,* 1980]. Fibrillar and granular components can also be distinguished in the multiple nucleoli or growing oocytes in the zebra fish [*Baumeister,* 1976]. Later on in stage 4 each extranucleolus appears as a ring carrying nucleolar granules. The same structure is seen in stage 5. In the oogonia and oocytes at the stage of early diplotene in silver carp, the granular component of nucleoli generally predominates over the fibrillar one [*Emel'Yanova,* 1979]. *Raikova* [1976] observed that in *Acipenser ruthenus* and *A. guldenstadti,* oogonial nuclei show single nucleoli and paranucleolar fibrillar masses which appear to be extruded from the nucleus to form specific proteinaceous bodies in the cytoplasm.

Although the shape and size of nucleolus is highly variable in different fish species and (or) stages of oocyte growth (fig. 1a–f, 2a–c, 11a–f), a relationship exists between a reticulate nucleolus and other signs of great activity in fish oocyte. The nucleolus generally becomes reticulate at the onset of vitellogenesis. But is becomes more roundish and compact in later stages.

The fibrous core and granular cortex demonstrated in nucleoli of oogonia and leptotene-pachytene oocytes in the viviparous teleost *(X. helleri)* form a typical configuration (fig. 19a, b) [*Azevedo and Coimbra,* 1980] that has equally been described in spheroidal nucleoli of immature amphibian oocytes [*Miller and Beatty,* 1969; *MacGregor,* 1972]. In the amphibians, these contain a DNA segment with spaced transcriptional units for the synthesis of 18S and 28S ribosomal RNA [*Miller and Beatty,* 1969]. The presence of nucleolar associated chromatin only during leptotene-pachytene in the fish oocytes (fig. 19c) [*Baumeister,* 1976; *Raikova,* 1976; *Azevedo and Coimbra,* 1980; *Ionescu,* 1980] bears a certain parallelism with the exclusive pachytene occurrence of chromatin caps serving for DNA amplification in amphibians [*MacGregor,* 1972; *Van Gansen and Schram,* 1972; *Ficq,* 1976]. *Chmilevskii and Raikova* [1976] observed that the pattern of

---

Fig. 17. Part of nucleolar body with interstices (I) showing particulate and fibrous components in the oocyte of lungfish. × 50,000 [from *Scharrer and Würzelmann,* 1969a].

Fig. 18. Two nucleolar bodies, one of them with multiple cavities, being in the process of fragmentation and dispersion in the oocytes of lungfish. × 27,000 [from *Scharrer and Würzelmann,* 1969a].

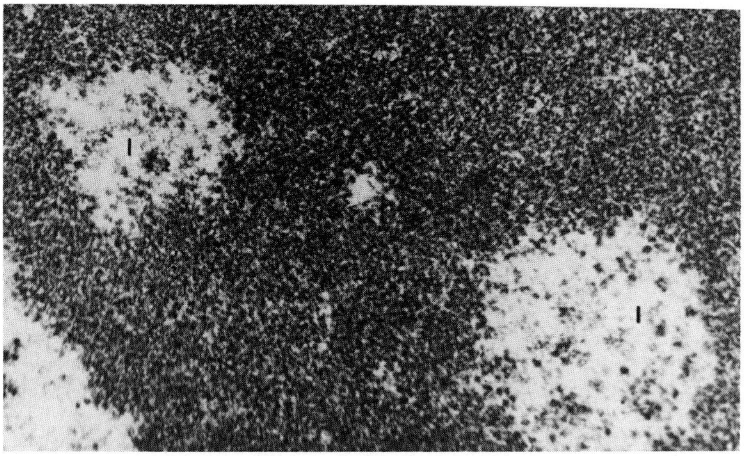

17

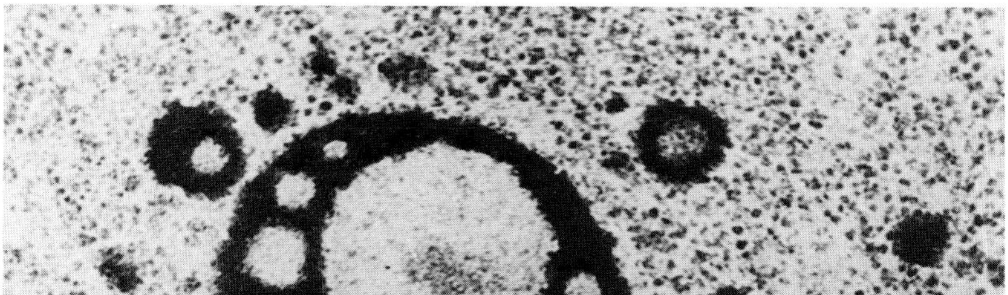

18

[^{3}H]-thymidine incorporation in meiotic oocytes of sterlet *(A. ruthens)* is similar to that in oocytes of Amphibia and Osteichthyes. In the oogonia as well as in the leptotene and zygotene oocytes, the label is predominantly localized over chromosomes. An intensive incorporation of [^{3}H]-thymidine into the material of the heterochromatic cap is seen during pachytene. These observations suggest that the main synthesis of extra DNA in the sterlet oocytes occurs during pachytene and no DNA is synthesized by the diplotene oocytes.

Raikova et al. [1977] made a study of localization of amplified ribosomal DNA in the meiotic prophase of acipenserid oocytes in situ hybridization with ^{125}I-ribosomal RNA. For these investigations, DNA in sections of sterlet *(A. ruthenus)* and sturgeon *(A. guldenstadti)* ovaries was hydridized with rRNA isolated from *Xenopus laevis* oocytes and labelled with ^{125}I. Hybridization results in specific labelling of the caps of extrachromosomal DNA which appear at the pachytene stage and at the early diplotene stage of meiotic oocyte nuclei. All the other cells such as oogonia, leptotene and zygotene oocytes, follicle cells, etc., bind no labelled rRNA. These observations have further confirmed that it is rDNA which is amplified in early oogenesis of the Acipenseridae; the amplified rDNA form a compact cap in pachytene oocyte nuclei. *Raikova* [1976] also confirmed that during pachytene, an extrachromosomal DNA cap is formed in the oocyte nucleus of *A. ruthenus* and *A. guldenstadti.* In diplotene, its material spreads beneath the nuclear envelope and gives rise to peripheral nucleoli which develop into amphinucleoli. Peripheral nucleoli are accompanied by proteinaceous nuclear bodies. During vitellogenesis, the peripheral nucleoli pull into radially arranged threads, then condense again. Evolution of acipenserid nucleoli during oogenesis resembles that of amphibians [*MacGregor,* 1972].

Since there does not occur subsequent multiplication of nucleoli during oogenesis of teleost *X. helleri* [*Azevedo and Coimbra,* 1980], it may be possible that if any DNA amplification takes place here, as shown for other teleosts [*Mazabraud* et al., 1975; *Denis and Wegnez,* 1977; *Ionescu,* 1980], all new DNA will have to fit within the growing 1 or 2 nucleoli of each nucleus [*Azevedo and Coimbra,* 1980].

Cytochemical techniques have demonstrated RNA and protein in the nucleoli of fish oogenesis [*Guraya,* 1963, 1965]. Abundant granular substance of the nucleoplasm also stains intensely for RNA and protein. *Azevedo and Coimbra* [1980] observed that 4 h after labelling with the [^{3}H]-uridine, both fibrillar and granular components of nucleolus in *X. helleri* are

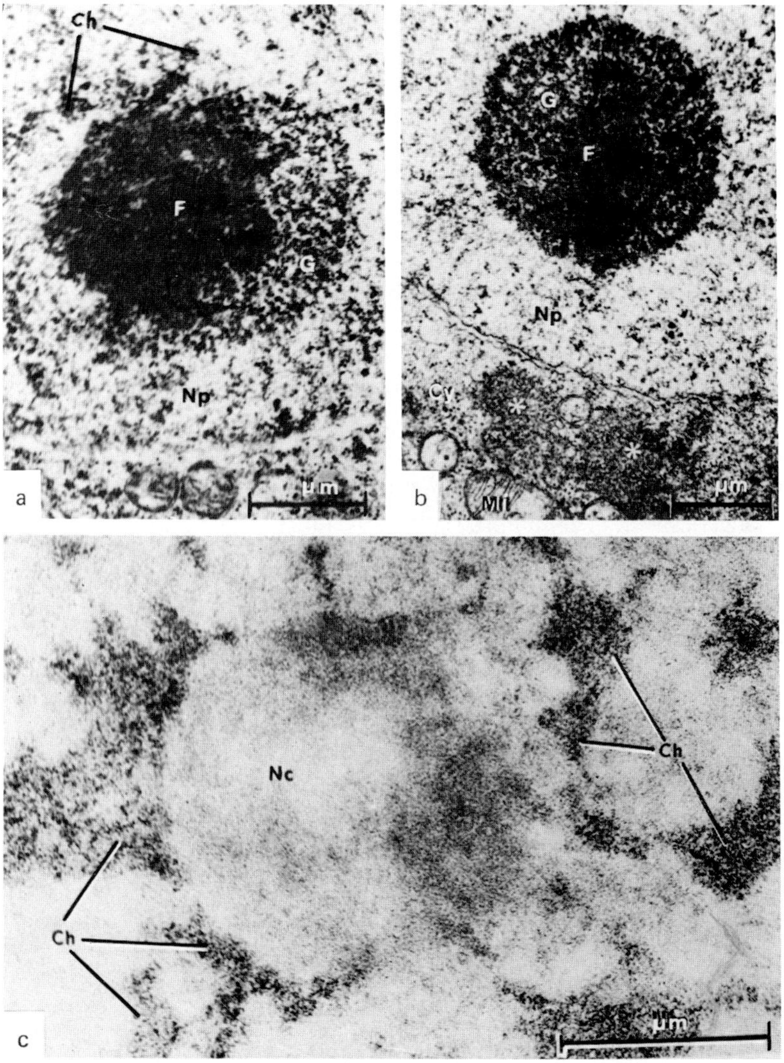

Fig. 19. From the oocytes of *Xiphophorus. a, b* Diplotene oocytes no longer close to the nuclear envelope. *a* At early diplotene, remains of perinucleolar chromatin (Ch) are visible; the fibrillar core is eccentric (F). *b* At a later time, there is no associated chromatin. In cytoplasm (Cy), larger perinuclear dense masses (*) close to mitochondria (Mi). *c* Zygotene. Osmium amine stain. Several DNA-containing chromatin masses (Ch) begin to surround the nucleolus (Nc). *a* × 27,000; *b* × 23,000; *c* × 48,000 [from *Azevedo and Coimbra,* 1980).

labelled at all meiotic stages. Labelling is increased at the hyperplasia of dictyate (ring-shaped nucleoli), but falls abruptly during the ensuing vitellogenesis. These observations indicate that a higher synthesis and export of nucleolar material consisting of RNA and protein, occur in the course of dictyate, peaking when ring-shaped nucleoli are formed. *Dutt* et al. [1975] demonstrated histochemically alkaline and acid phosphatases in the nucleoli of *Anabas scandens.* These phosphatases may be playing some roles in the metabolism of nucleoli, which need to be determined at the molecular level. The nucleolus of early oocytes also contains phospholipids [*Gopal Dutt and Govindan,* 1967].

2. Nucleolar Vacuoles and Their Significance

The large nucleoli placed close to the nuclear envelope of large previtellogenic and early vitellogenic oocytes develop vacuoles of variable size (fig. 13–15) as also stated already. Ultrastructural evolution of nucleoli during oogenesis in *Xiphophorus helleri* is illustrated in figure 15. At early dictyate, they increase in size from 6 to 18 μm and acquire an increasing number of vacuoles, becoming ring-shaped by mid-dictyate (fig. 20a–d). Abundant dense material occurs at this stage on both sides of the numerous nuclear pores which develop during the oocyte growth (fig. 20e). During late dictyate, while vitellogenesis takes place in the ooplasm, the nucleoli are again multivacuolated and ultimately compacted. The size of nucleoli is decreased markedly and they disappear when the maturing oocyte is ready for fertilization.

While in *Xenopus laevis* and acipenserid oocytes nucleolar growth occurs entirely during diplotene [*Van Gansen and Schram,* 1972; *Raikova,* 1976], in the viviparous fish *X. helleri,* size increase of nucleolus begins at diplotene but is only marked at dictyate [*Azevedo and Coimbra,* 1980]. The simultaneous development of vacoules in the nucleolar mass appears to be a generalized phenomenon in animal oogenesis [see also *Raven,* 1961; *Scharrer and Würzelmann,* 1969a; *Baumeister,* 1976; *Adiyodi and Adiyodi,* 1983], for exceeding the domain of invertebrates and lower vertebrate species since it has also been reported in birds and mammals [*Mirre and Stahl,* 1976; *Palombi and Viron,* 1977].

In some cell types, the presence of vacuoles in the nucleolus has been equated with a high rate of RNA synthesis or with a surge of cytoplasmic migration of nucleolar material [*Azevedo and Coimbra,* 1980]. *Azevedo and Coimbra* [1980], however, believe that vacuolization of nucleoli during the oogenesis of viviparous teleost *X. helleri* may be the result of an increased

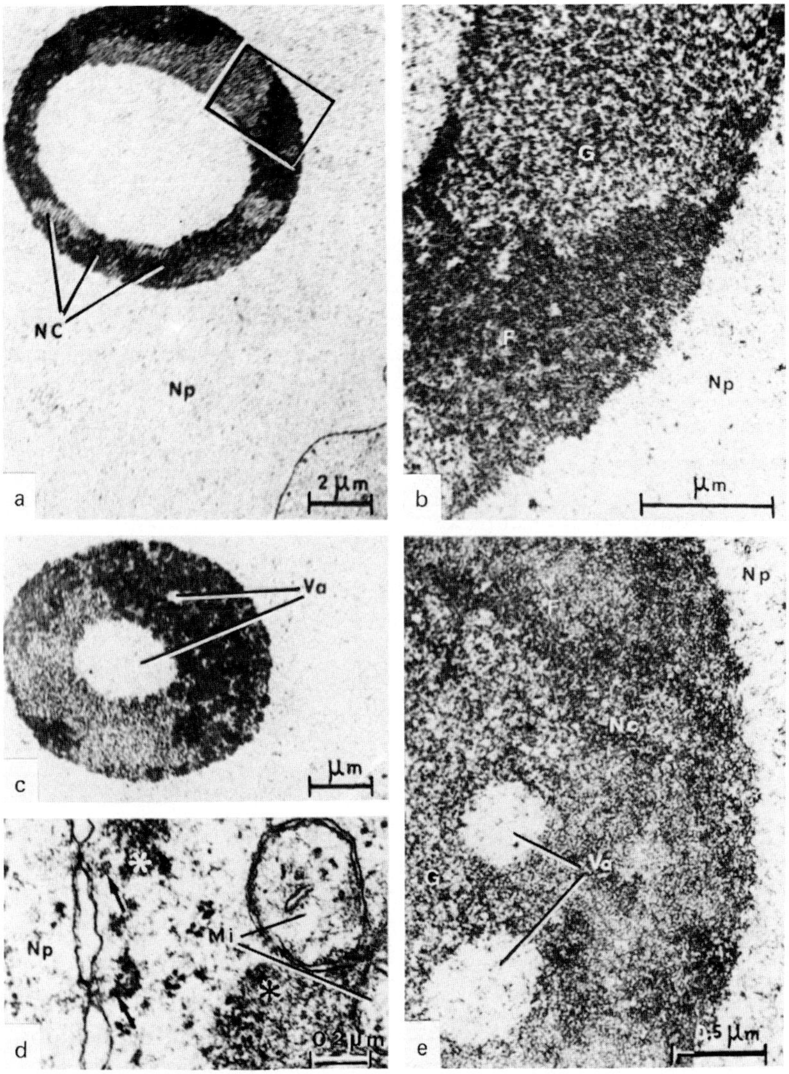

Fig. 20. From the oocytes of *Xiphophorus.* *a* Ring-shaped nucleolus from mid-dictyate. Rectangle is enlarged in *b.* *b* Detail of *a,* showing the marked separation of fibrillar (F) and granular (G) components. *c* Another ring-shaped nucleolus at a slightly later period of dictyate showing a new vacuole in addition to the central large one. *d* Detail of the juxtanuclear region of a mid-dictyate oocyte. Note dense material at the nuclear pores (arrows) and a perinuclear dense mass (*) surrounded by the mitochondria. *e* Detail of a late dictyate nucleolus with vacuoles in lieu of central cavity and a certain intermingling of fibrillar (F) and granular (G) components. *a* × 6,500; *b* × 30,000; *c* × 11,000; *d* × 65,000; *e* × 42,000 [from *Azevedo and Coimbra,* 1980].

synthesis and exportation of nucleolar materials. Ring-shaped nucleoli show a larger silver grain density per unit area of nucleolar mass than earlier compact forms at diplotene. Furthermore, and in spite of the large size of central vacuole (1,150 μm^3), the ring itself shows a much larger volume (900 μm^3) than that of the largest diplotene compact oocytes (113 μm^3). This suggests that highly vacuolated nucleoli being much larger and more active in RNA synthesis certainly form a greater bulk of RNA than earlier compact forms. But this suggestion needs to be supported by results obtained with recombinant DNA methods. On the other hand, a considerable transit of nucleolar material towards the cytoplasm is suggested by the presence of nucleolus-like dense masses on both sides of nuclear pores (fig. 4a–d, 16, 20e) as well as in the nucleoplasm. The relatively high grain density over nucleolar vacoules is surprising and may be due to some possible accumulation of recently synthesized material [*Azevedo and Coimbra,* 1980]. Since only one experimental period (4 h) was studied quantitatively by *Azevedo and Coimbra* [1980], these results need to be extended and confirmed from pulse labellings. Fibrillar centres [*Recher* et al., 1969] have not been observed. All vacuoles, even those with a fine flocculent background (fig. 20a–d), are much less electron dense than the fibrillar centres and not entirely surrounded by the fibrillar component.

3. Origin

In the light microscope studies, the extra nucleoli during the fish oogenesis were believed to be formed by division or fragmentation of the original nucleoli [*Raven,* 1961; *Khanna and Pant,* 1967; *Shrestha,* 1980]. However, according to the modern view, the nucleoli originate from certain heterochromatic regions of chromosomes known as nucleolus organizers [*Vincent,* 1955; *K. Yamamoto,* 1956b; *Arndt,* 1960a, b; *Chouinard,* 1963; *Rastogi,* 1968a; *MacGregor,* 1972; *Baumeister,* 1976]. A nonheterochromatic region distinct from the nucleolar organizing region is also instrumental in the formation of extra nucleoli [*Chouinard,* 1963]. In most amphibians, the morphological evolution of the nucleolus during oogenesis is characterized by three principal events [*Callan,* 1966; *MacGregor,* 1972]. These include (a) the establishment of specific relationships with the nucleolar organizer at pachytene; (b) the striking growth and/or multiplication of the nucleolar mass of diplotene with the occurrence of vacuolizations and ring-shaped configurations, and (c) the regression and disappearance of nucleoli when vitellogenesis has taken place. Most biochemical phenomena underlying such alterations are well worked out for their elucidation, especially in amphibians [*MacGregor,* 1972]. But such biochemical phenomena are relatively less worked out for fish. However, *Denis* [1977] observed that the oocytes of teleosts and amphibians multiply their genes 28S and 18S from the start of their growth. But these genes remain inactive during the first phase of oogenesis. During this period the oocyte accumulates most exclusively RNA 5S and transfer RNA. The 28S and 18S genes become active at the beginning of real growth. Recombinant DNA methods will be very useful to follow the precise nature and regulation of activity of genes involved in the formation of nucleoli (or ribosomal RNA) in the fish oocyte which accumulates abundant rRNA.

4. Function

Various studies have revealed that the chromatin caps embracing the nucleoli at pachytene contain amplified copies of DNA coding for ribosomal 18S and 28S [*McGregor,* 1972; *Ficq,* 1976; *Denis,* 1977], which are distributed during the ensuing diplotene into multiple nucleoli as already described. The latter are now the established sites of synthesis of ribosomal RNA in the fish oogenesis [*Baumeister,* 1973, 1976]. Nucleolar incorporation of RNA precursors is maximum during stage 3 of oocyte growth in the zebra fish [*Baumeister,* 1976]. The rapid labelling is localized in the fibrillar part of the peripheral nucleoli which contribute more than 60% of the RNA synthesis of an oocyte nucleus. A massive transport of macromolecular RNA into the ooplasm is observed in previtellogenic oocytes but ceases in stage 4. This is accompanied by the appearance of densely staining layers on the nuclear envelope. Most of the RNA synthesized late in oogenesis is stored within the nucleus for embryonic protein synthesis.

Ribosomal RNA synthesized in the nucleoli accumulates in the large number of ribosomes that appear in the ooplasm before yolk deposition [*Denis,* 1977]. The DNA of the nucleolar core has the base sequence needed for transcription of ribosomal RNA. RNA is transcribed, complexes with protein and eventually distributed between 50S and 30S ribonucleoprotein particles [*MacGregor,* 1972]. Regulation of synthesis of abundant ribosomal proteins must be determined in future studies by applying recombinant DNA technology to fish oogenesis. Without the synthesis of these proteins, no ribosomes can be formed. Besides the enormous synthesis and storage of 18S and 28S RNA at diplotene, oocytes snythesize 5S RNA and transfer RNA during the whole course of oogenesis. These products accumulate in 8S and 42S RNP cytoplasmic particles preceding the integration of 5S RNA in the ribosomes [*Denis,* 1977]. Recent biochemical studies also suggest that the nucleolus plays an important role in the relay of information via RNA from the nucleus to the cytoplasm in the amphibian oogenesis. Similar biochemical changes have also been reported for the teleost oogenesis, namely amplification of genes coding for 18S and 28S RNA and the continuous formation of 5S RNA and transfer RNA [*Mazabraud* et al., 1975; *Denis and Wegnez,* 1977].

Chaudhuri and Mandal [1980] studied the accumulation of DNA, RNA and protein in developing oocytes of catfish *(Heteropneustes fossilis).* While DNA content per oocyte increases only severalfold during development of small oocytes to mature eggs, RNA and protein content increases by about 2 and 3 orders of magnitude, respectively; the increase in DNA content may be related to amplification of genes. The accumulation of RNA and protein is very rapid in mid-vitellogenic stage oocytes. This is also in agreement with the histochemical observations which have revealed an abundant basophilic substance consisting of RNA and protein in the nucleoplasm and ooplasm of such oocytes [*Guraya,* 1965]. Small oocytes contain only small amounts of rRNA. With the oocyte growth, RNA accumulation takes place so that in unfertilized mature eggs, the major fraction is rRNA. The isolated oocytes show the presence of high Mr rRNA and their precursor, but very little transfer RNA. Regulation of formation of stage-specific RNA species in fish oogenesis is still not known and thus must be determined in future studies; even the proteins of the germinal vesicle need to be studied.

Wegnez et al. [1978] suggested that four biochemical mechanisms operate in the oocytes of teleosts and amphibians: (1) amplification of the 28S and 18S genes; (2) noncoordinate accumulation of 5S RNA and 28S + 18S RNA; (3) storage of 5S and transfer RNA made in excess by small oocytes within nucleoprotein particles, and (4) expression of different 5S genes in oocytes and somatic cells. These workers further observed that ribosomal gene amplification is possibly low or absent in the oocytes of the dogfish *Scyliorhinus canicula.* Its previtellogenic oocytes accumulate more 5S RNA than needed for ribosome assembly. Transfer and 5S RNA present in small oocytes are probably not free in the cell sap. A substantial fraction of these RNAs sediments at 10S when homogenates of immature ovaries are centrifuged in sucrose density gradients. In contrast to what has been observed in teleosts and amphibians, 5S RNA from ovaries of *S. canicula* are identical in sequence to 5S RNA from liver. Among the 4 mechanisms described above for the teleosts and amphibians, the 2nd and possibly the 3rd are used by the oocytes of *S. canicula.* Mechanism (4) is absent in this species. In conclusion, it can be stated that further studies are needed to elucidate the regulatory mechanisms involved in transcription of various RNA species during the oocyte growth in fish.

D. Intranuclear Lipid Bodies and Inclusions

1. Lipid bodies

Histochemical studies have demonstrated the presence of sudanophilic lipid bodies in the nucleus of previtellogenic oocytes in teleost ovary [*Subramaniam and Aiyar,* 1935; *Seshachar and Nayyar,* 1963; *Guraya,* 1965; *Gopal Dutt and Govindan,*1967; *Saxena and Bhatia,* 1977; *Ramadan* et al., 1979b]. *Seshachar and Nayyar* [1963] described sudanophilic lipid bodies in the nucleus of previtellogenic oocytes of *Heteropneustes fossilis,* which consist of phospholipids and triglycerides as also reported by *Saxena and Bhatia* [1977]. According to *Gopal Dutt and Govindan* [1967], the intranuclear lipids composed of triglycerides appear in the nucleus of early oocytes of *Anabas scandens* but disappear during subsequent stages of oocyte growth. Sudanophilic lipid bodies of heterogeneous appearance are not only seen in the developing oocyte nucleus of the teleost *(Channa marulius)* (fig. 21a–c, 22a, b) but also in that of the toad *(Bufo stomaticus)* [*Guraya,* 1965]. These lipid bodies consisting of phospholipids either lie singly or in groups and do not show any particular anatomical association with nucleoli of both fish and toad oocytes. They are sparsely distributed in the central portions of the nucleus as well as adjacent to the nuclear envelope (fig. 21a–c, 22a, b). Their number and distribution vary in different oocytes in the same range of size. With the increase in the size of nucleus their number is also increased.

Guraya [1965] observed that with the origin of yolk, all the lipid bodies of the nucleus begin to move close to the nuclear envelope in the oocytes of both the fish and toad (fig. 23b). Meanwhile, they begin to appear outside the folded nuclear envelope but still lie very close to the latter. But there is no morphological evidence of their extrusion into the ooplasm at the light microscope level. The lipid bodies appear to arise de novo outside the nuclear envelope as they disappear from inside of the nucleus. Their material may be diffusing outside the nucleus to constitute new lipid bodies which again originate in the form of groups of granules. The latter further grow and coalesce to form one large body of heterogeneous appearance. Some of these lipid bodies may also be associated with a vacuole-like space

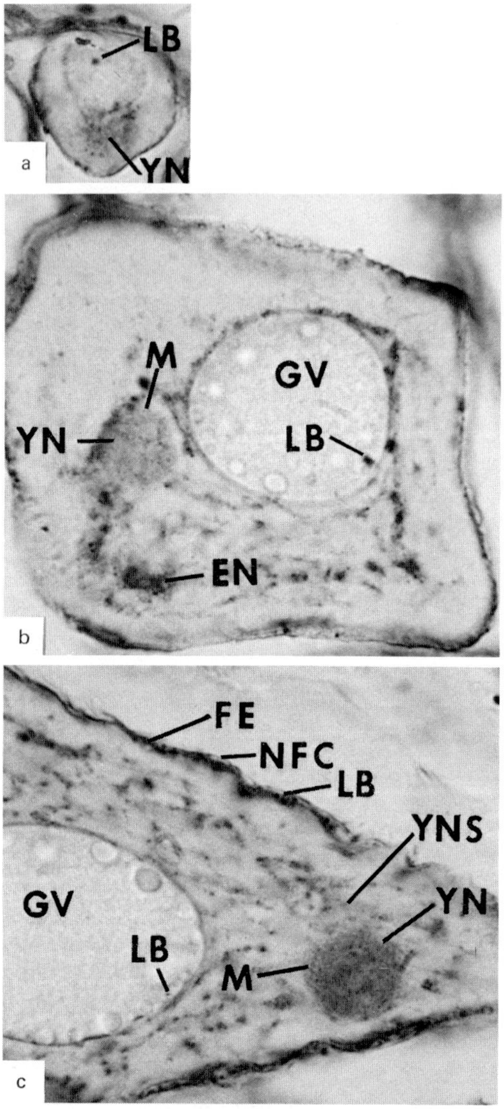

Fig. 21 a–c. Photomicrographs of oocytes of *Channa* fixed in formaldehyde-calcium and postchromed in dichromate-calcium and their frozen gelatine sections were colored in Sudan black B, showing lipid bodies (LB) and nucleoli (sudanophobic) in the germinal vesicle (GV) and yolk nucleus (YN), irregular masses of yolk nucleus substance (YNS), mitochondria (M), lipid bodies, and extruded nucleolus (EN) in the cytoplasm. Lipid bodies (LB) are also seen among the small follicle cells (NFC) of the follicular epithelium (FE) which is very poorly developed. × 1,480 [from *Guraya,* 1965].

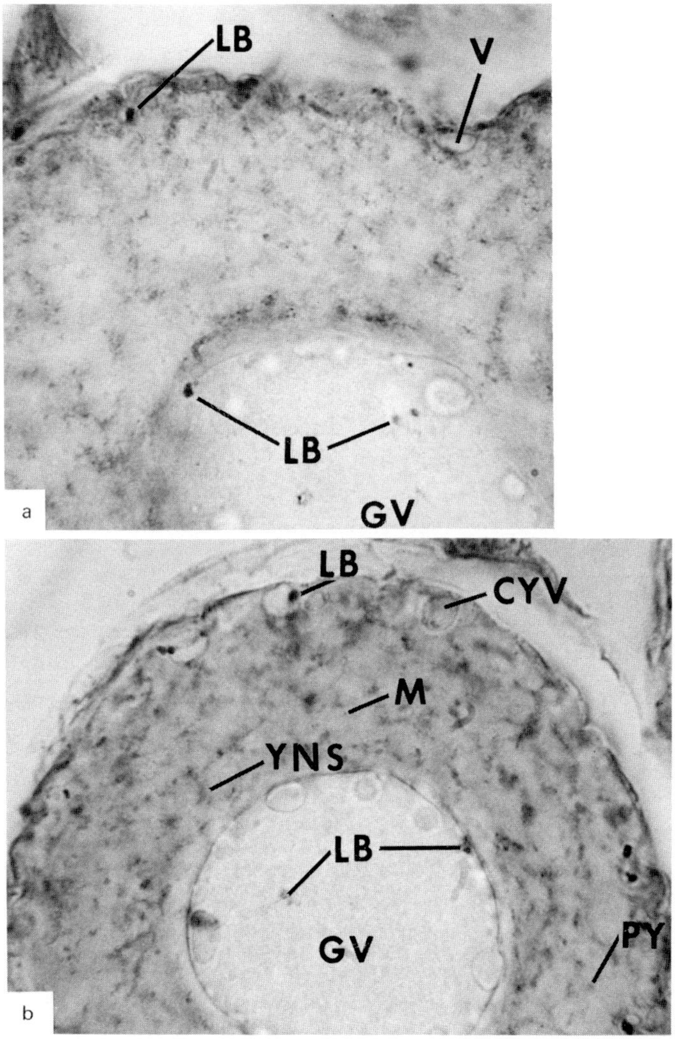

Fig. 22a, b. Photomicrographs of portions of early vitellogenic oocytes of *Channa,* prepared as in figure 21 demonstrating the appearance of sudanophobic vacuoles (V) and cortical alveoli (CYV) associated with lipid bodies adjacent to the plasma membrane of the oocytes. The proteid yolk granules (PY) originate in association with the yolk nucleus substance (YNS) and granular mitochondria (M). The germinal vesicle (GV) shows some lipid bodies (LB) and nucleoli situated close to the nuclear membrane. Some lipid bodies, at this stage, lie outside the nuclear envelope as seen in figure 24a. × 1,480 [from *Guraya,* 1965].

formed by the evagination of the nuclear envelope. Such associations are similar to peripheral pino- and phagocytotic vacuoles and lipid bodies formed adjacent to the plasma membrane of growing previtellogenic oocytes as will be described in Chapter III. The lipid bodies consisting of phospholipids continue to be visible outside the lobulated nuclear envelope during vitellogenesis. When the yolk is formed, the lipid bodies and vacuoles disappear in situ close to the nuclear envelope. *Ramadan* et al. [1979b] have also reported sudanophilic lipid bodies in the nucleus of young oocytes in *Merluccius merluccius*, which finally pass into the ooplasm to play some role in yolk formation.

2. Inclusions

Electron microscope studies have demonstrated intranuclear inclusions in the oocytes of viviparous teleost *(Xiphophorus helleri)* during prophase I [*Azevedo and Coimbra*, 1976; *Azevedo*, 1976]. In osmium-fixed leptotene nuclei, the inclusions consist of groups of membrane-limited vesicles and tubules with pale contents, placed near the inner nuclear membrane with which some of them show apparent continuities. In zygotene nuclei, larger vesicles also appear to be bounded by two or three membranes and having tubules apparently invaginated from their walls (fig. 24a, b). In pachytene-dictyate nuclei most vesicular bodies show a wall formed by stratified membranes, or are entirely composed of membranous whorls (fig. 25a, b). In glutaraldehyde-osmium fixed material some of these myelin-like bodies show a peculiar arrangement consisting of concentric bands each containing thick inner dense lamellae 2.0–3.0 nm thick and a 5.0 nm outer lamellae. These inclusion bodies are suggested to originate from the inner nuclear membrane of oocytes when cells begin to grow intensely during prophase I.

The nuclear bodies in the growing oocytes of *X. helleri* appear to result from an exaggerated production of perinuclear membrane. The intranuclear bodies appear to become more complex at late prophase (fig. 26a, b), probably by association of individual vesicles and the occurrence of multiple membrane invaginations, which may be related to active metabolic phenomena taking place at this stage in oocytes. These complex intranuclear inclusion bodies (fig. 25, 26) appear to correspond to intranuclear lipid bodies of light microscopy (fig. 21, 22), which consist either of phospholipids or phospholipids and some triglycerides as already described. The physiological significance of these intranuclear lipid bodies or inclusions needs to be determined with correlative electron microscopic and autoradio-

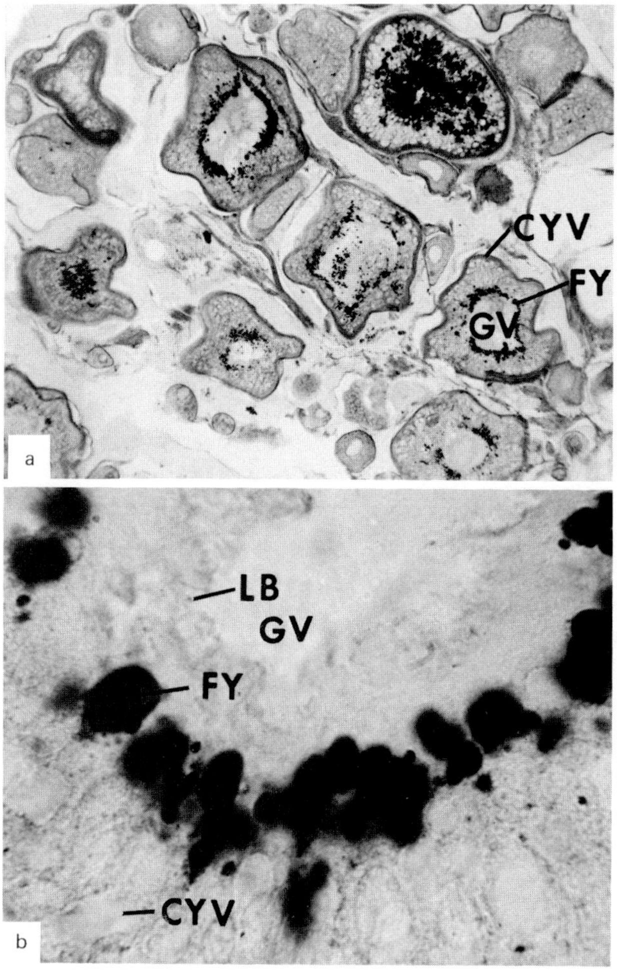

Fig. 23a, b. Photomicrographs of ovary and perinuclear zone of large oocyte of *Channa* respectively, prepared as figure 21, showing various stages of previtellogenesis and vitellogenesis as revealed by the fatty yolk (FY) and cortical alveoli (CYV). The germinal vesicle (GV) is sudanophobic as compared with the cytoplasm which is lightly sudanophilic due to the presence of mitochondria and yolk nucleus substance. *b* Details of deposition of fatty yolk (FY) in association with the homogeneously distributed yolk nucleus substance of the perinuclear zone. Small lipid bodies (LB) are seen outside the lobulated nuclear envelope of the germinal vesicle (GV), *a* × 47; *b* × 810 [from *Guraya,* 1965].

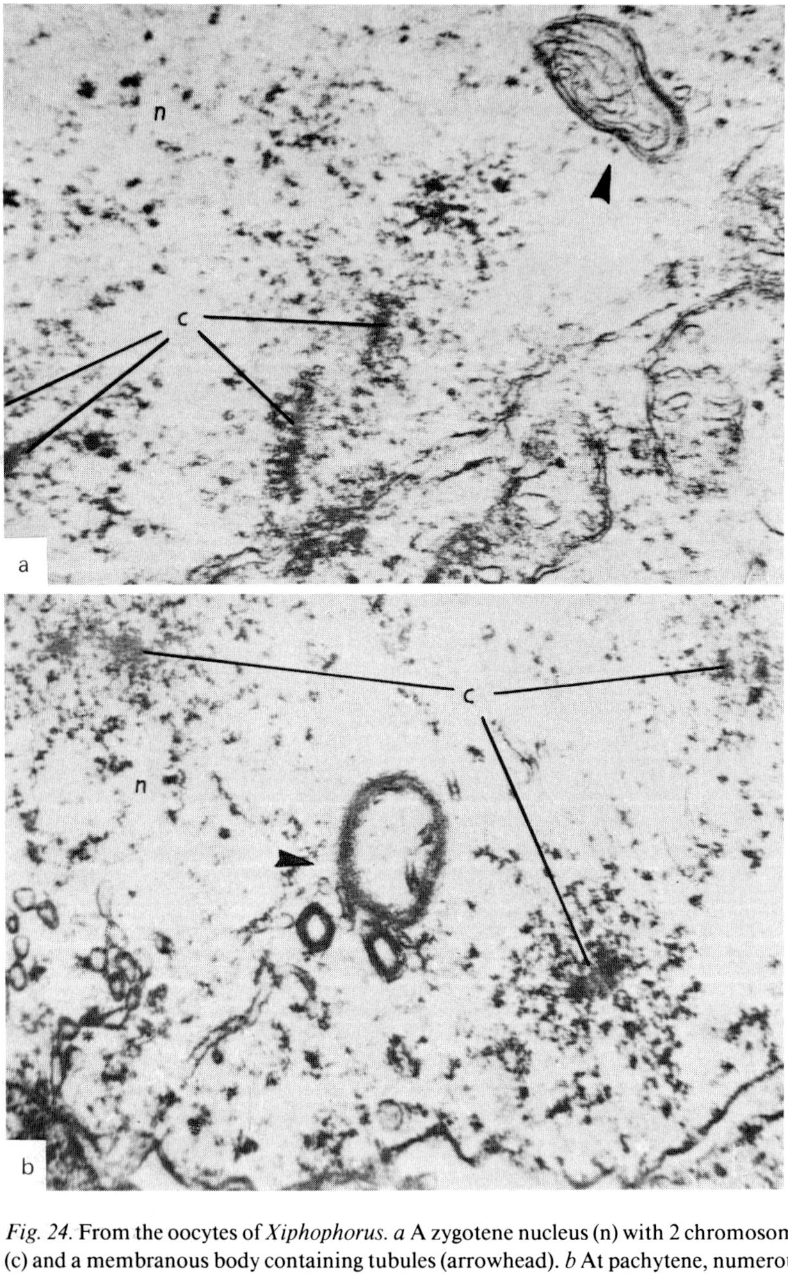

Fig. 24. From the oocytes of *Xiphophorus*. *a* A zygotene nucleus (n) with 2 chromosome pairs (c) and a membranous body containing tubules (arrowhead). *b* At pachytene, numerous simple vesicles and tubules close to the inner nuclear membrane (*) and 3 vesicles (arrowhead) with 2 or 3 membranes at the wall and a tubule in the interior of the larger vesicle. Of the 3 pairs of chromosomes (c) the synaptonemic complex is evident in the one situated on the right. Osmium fixation. *a* × 54,000; *b* × 37,800 [from *Azevedo*, 1976].

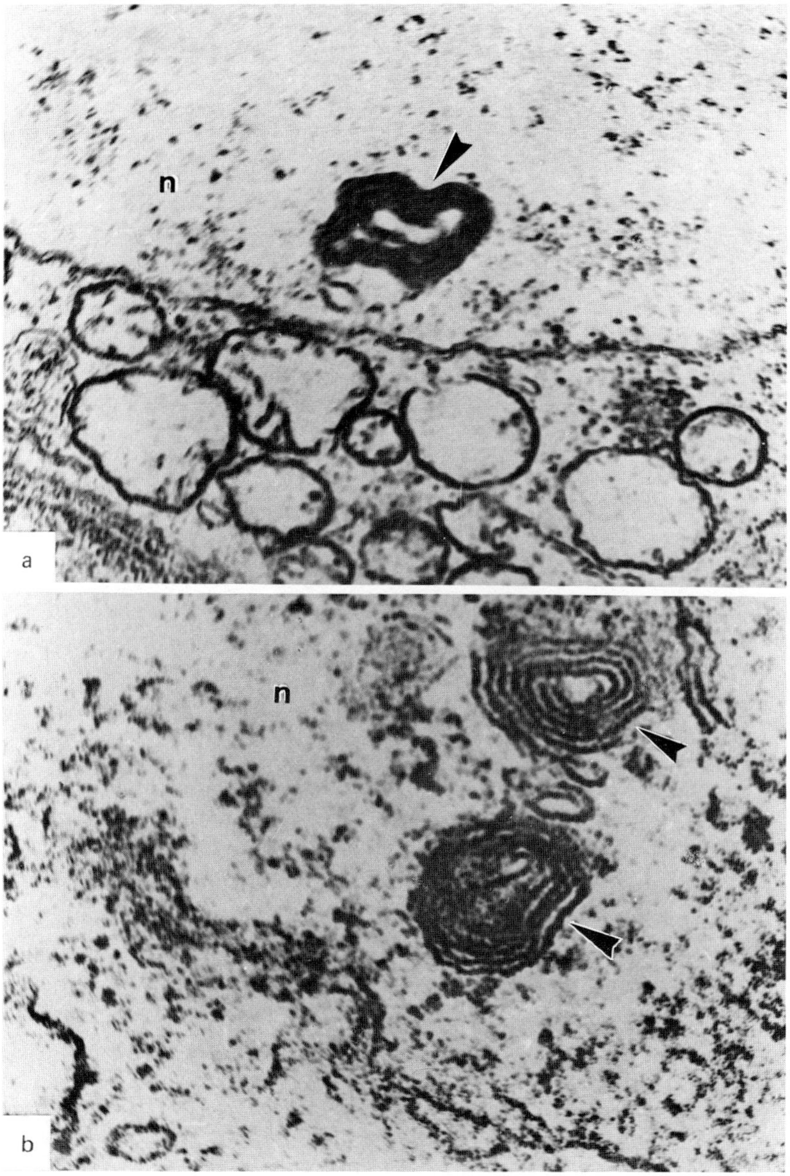

Fig. 25. From the oocytes of *Xiphophorus. a* In the (upper) oocyte nucleus (n) at pachytene, a body with stratified membranous wall can be seen (arrowhead). The nucleus (n) at the lower left belongs to a follicular cell. *b* Two intranuclear bodies (arrowheads) made up of membranous whorls. Note the unit membrane structure of the latter. Osmium fixation. *a* × 36,000; *b* × 120,000 [from *Azevedo,* 1979].

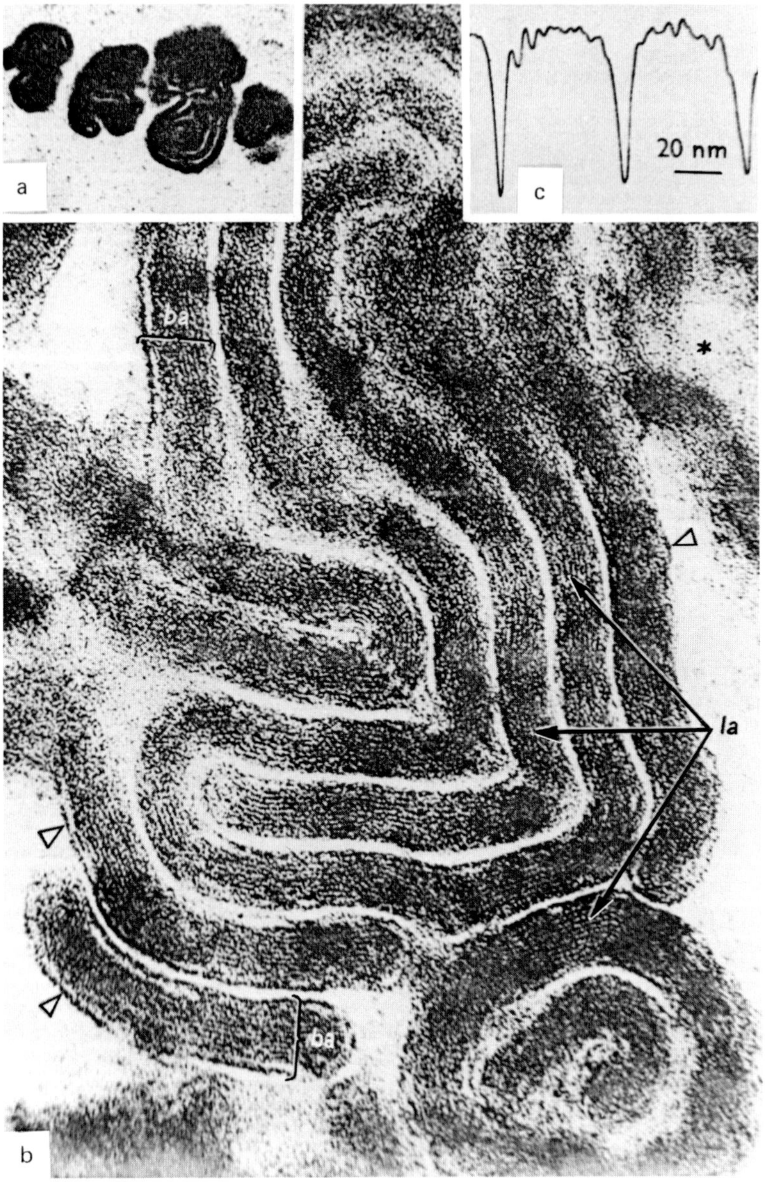

Fig. 26. From the oocytes of *Xiphophorus. a* Three inclusion bodies in a dictyate nucleus formed by concentric dense bands. *b* High magnification of the same type of bodies depicted in *a,* showing the wide bands (ba) composed of tightly dense lines (la) the outer ones of which are thicker (hollow arrowheads). A grazing section shows blurred membrane profiles (*). *c* Densitometric reading of the bands, each small peak corresponding to a dense line. Glutaraldehyde-osmium fixation. *a* × 33,000; *b* × 225,000 [from *Azevedo,* 1976].

graphic studies. Their development in the nucleus of fish oocyte may be related to its peculiar metabolism as evidenced by the extensive development of nucleolar apparatus which is involved in the production of abundant ribonucleoproteins for the ooplasm as already discussed. The lipids of inclusion bodies may be playing some significant roles in the biology of the enlarged nucleus.

E. Nucleocytoplasmic Relations and Nucleolar Extrusions

In various studies on the interactions between nucleus and cytoplasm in the oocytes of various animal species, much emphasis is laid on the nucleolus because of its significant role in the biogenesis of cytoplasmic ribosomes [*MacGregor,* 1972; *Wischnitzer,* 1973; *Davidson,* 1976]. Ribosomal RNAs formed in the nucleolar apparatus, apparently by way of precursor molecules [*Rogers,* 1968; *MacGregor,* 1972; *Wischnitzer,* 1973; *Davidson,* 1976], need transport mechanisms by which material of this type may reach the cytoplasm. Traffic across the nuclear envelope can, therefore, be expected to be particularly active during those early stages of oocyte growth when excessive amounts of ribosomes are formed and stored for future use during embryonic development [*Davidson,* 1976]. Possible mechanisms for nuclear-cytoplasmic transport are diffusion of nonparticulate elements, transport of small nucleolar fragments across pores of the nuclear envelope, or migration of nucleolus/nucleoli as such across the nuclear envelope or development of larger nuclear blebs whereby a segment of the entire nuclear envelope or its outer membrane encloses the material to be transported into the cytoplasm and may persist for sometime in the cytoplasm [*Raven,* 1961; *Scharrer and Würzelmann,* 1969a; *Wischnitzer,* 1973]. Recent correlative electron microscopic and biochemical including autoradiographic studies have suggested the presence of following possible mechanisms for the transport of nucleolar material in the fish oogenesis. Factors of these mechanisms of transport need to be resolved in future studies.

1. Formation of Blebs

Scharrer and Würzelmann [1969a] observed that numerous nucleolus-like bodies form a close spatial relationship with the nuclear envelope by extending pedicels and streams of finely dispersed material towards the inner membrane of nuclear envelope in the oocytes of the African lungfish

(Protopterus aethiopicus) (fig. 13). An additional relevant characteristic is the spacing of nuclear pores and blebs in relation to these sites of contact with nucleoli (fig. 4a–d, 7a, b). At such contact points, gaps in the perinuclear cistern are of more common occurrence than elsewhere along the nuclear boundary. Evaginations of the cuter nuclear membrane form blebs with or without visible content (fig. 4a, 7a, b) and these are finally cut off to form small vesicles in the perinuclear cytoplasm [see also electron micrographs in *Riehl,* 1976a, 1978a; *Upadhyay,* 1977].

2. Nuclear Pores

Small, electron-dense aggregates, indistinguishable from nucleolar material, are seen on both sides of the nuclear envelope opposite to each other, some being connected by a slender portion of the same material within a nuclear pore (fig. 3b, 4a–d, 5, 15, 20e) [*Droller and Roth,* 1966; *Scharrer and Würzelmann,* 1969a, b; *Riehl,* 1976a, 1978a; *Azevedo and Coimbra,* 1980; *Monaco* et al., 1981]. Such accumulations are believed to be formed from the detached parts of nucleolar bodies in transit to ooplasmic sites where they apparently participate in the biogenesis of ribosomes (fig. 18). This suggestion is supported by the fact that at the height of nucleolar emission, nucleoplasm and perinuclear ooplasm are so rich in small electron-dense particles of nucleolar origin that these are almost indistinguishable from each other in electron micrographs (fig. 3–7, 13, 18, 19a, b). Such configurations are generally interpreted as ultrastructural evidence for the presence of nucleolo-ooplasmic transport. This suggestion is further supported by the fact that in iron haematoxylin preparations the nucleoli and nucleolar material lying in the nucleoplasm and ooplasm show similar morphology and staining reactions (fig. 6a–e). The granular material of nucleolar origin, which consists of RNA and protein, forms conspicuous aggregations in the perinuclear cytoplasm of early vitellogenic oocytes (fig. 16b). Meanwhile, such material is also present in abundance in the nucleoplasm (fig. 16b, c). The morphology and distribution of nucleolar material both in the nucleus and ooplasm are indicative of their close relationship. These morphological observations obtained with light microscope are further supported by correlative cytochemical and electron microscopical studies on the nucleoproteins in the oogenesis of unisexual fish, *Poecilia formosa* [*Monaco* et al., 1981] and *Chrowis chrowis* [*Donato* et al., 1980]. Prominent accumulations of ribonucleoproteins (RNP) occur in nucleoli and appear on either side of the nuclear envelope during diplotene. In certain planes of section, RNP material appears to be in transit across this interface.

3. Direct Passage Across the Nuclear Envelope

At the stage of massive transport of nucleolar material, the route provided by the nuclear pores may be insufficient to cope with this type of situation and thus another, more spacious, gateway apparently begins to function. The latter involves direct passage of material across the nuclear membranes preferentially where these form blebs (fig. 7a, b) [*Scharrer and Würzelmann,* 1969a]. This suggestion gets support not only by the spatial relationships between nucleolar pedicels and blebs, but by the occurrence within perinuclear lacunae and blebs of particles that apparently derive from nucleolar bodies. Furthermore, frequent interruptions in the nuclear membranes preferentially placed where they expand into outpocketings indicate that at these sites temporary gateways may develop in the living oocyte that allow easy transport of intranuclear components to the ooplasm [*Riehl,* 1976a, 1978a]. This type of pathway is believed to be dictated not only by the amount of material to be transported at a given time but also by particle size. The nature and the physiological significance of the control mechanism involved need to be determined.

4. Nucleolar Extrusions and Their Function

In this mechanism of transport of nucleolar material, nucleoli as such or their large fragments are passed into the ooplasm. In the earlier literature, this phenomenon at the light microscope level was generally referred to as the nucleolar extrusions [*Raven,* 1961]. Actually, very divergent views were expressed about the presence or absence of nucleolar extrusions in the fish oogenesis [*Raven,* 1961]. Nucleolar extrusions were observed in some growing previtellogenic oocytes of teleosts such as *Channa marulius* [*Guraya,* 1965], *Mystus tengara* [*Guraya* et al., 1975], *C. striatus* [*Srivastava and Swarup,* 1979], *Merluccius merluccius* [*Ramadan* et al., 1978], and stone loach, *Neomacheilus beavani* [*Shrestha,* 1980]. According to *Guraya* [1965], the morphological and cytochemical changes noticed in the extruded nucleoli of *Channa* oocyte provide evidence for their actual extrusion from the nucleus. After the nucleolus passes into the ooplasm, it breaks up into pieces (fig. 27) and meanwhile it develops lipoproteins and some lipid granules in addition to its original RNA and proteins; lipoproteins demonstrable with histochemical techniques may be the result of some unmasking process as the intact nucleoli lying in the nucleus do not show such lipid component. However, nucleoli in general are known to contain some phospholipids demonstrated with biochemical techniques [see *Davidson,* 1976, for references]. Finally, the substrate of extruded nucleoli is dissolved in the

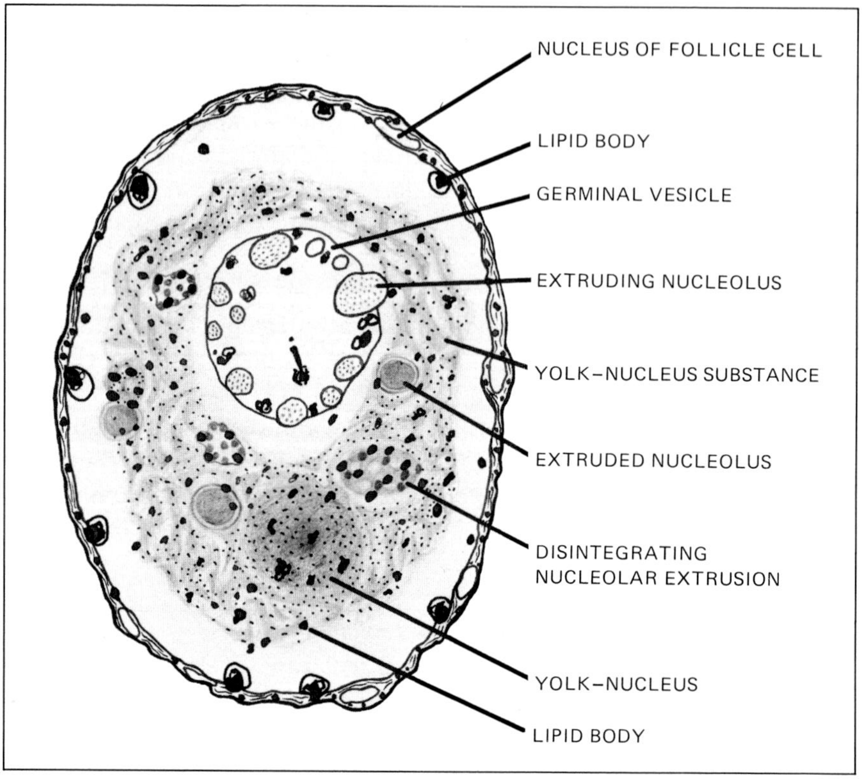

Fig. 27. Figure of previtellogenic oocyte of *Channa* illustrating various components of the germinal vesicle and cytoplasm [from *Guraya,* 1965].

cytoplasm. The dense bodies, described in the vicinity of yolk nucleus of *Channa* by *Guraya* [1963], are also the nucleolar extrusions.

The present author has made a more thorough study of nucleolar extrusions in the growing oocytes of several teleosts *(Colisa fasciatus, Notopterus notopterus, Cyprinus carpio)* which have been stained with iron haematoxylin [*S.S. Guraya,* unpubl. observations]. Nucleoli similar to those of nucleoplasm are seen in the ooplasm of previtellogenic and early vitellogenic oocytes (fig. 2b, c, 16a). Their extrusion into the ooplasm is supported by the fact that some of them are seen passing through the nuclear envelope (partly lying inside and partly outside the nuclear envelope) (fig. 1, 2, 11). The extruded nucleoli finally disintegrate and their material is distributed in the ooplasm. The nucleolar extrusions show many variations in their

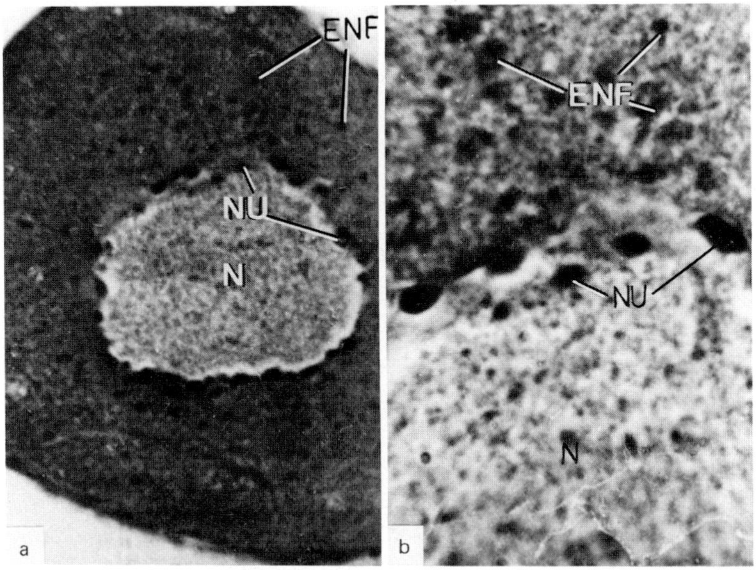

Fig. 28a, b. The extruded nucleoli and their fragments (ENF) in the ooplasm of very early vitellogenic oocytes of *Notopterus notopterus.* Nucleus (N) also shows nucleoli (NU) which are lying adjacent to the nuclear envelope. Iron-haematoxylin preparation. *a* × 300; *b* × 750.

occurrence in different stages of oocyte growth not only in the same species but also in different species (fig. 1, 2, 11, 16). *Kapoor* [1977] also observed that some of the nucleoli lying in contact with the nuclear envelope are passed into the surrounding cytoplasm of oocytes in *Puntius ticto.* The extruded nucleoli finally disappear near the periphery of the oocyte. *Ramadan* et al. [1978] have reported nucleolar extrusions during the yolk deposition stage of *M. merluccius.*

Besides the transport of nucleoli as such, there also occurs the passage of numerous nucleolar fragments into the ooplasm, which are mainly distributed in the perinuclear ooplasm (fig. 28) [*S.S. Guraya,* unpubl. observations]. Their transport is supported by the fact that the extruded nucleolar fragments closely resemble the nucleoli and their fragments lying in the nucleoplasm (fig. 28). *Saxena and Bhatia* [1977] observed that as the oocyte of *Heteropneustes fossilis* grows, the nucleoli increase in size and undergo fragmentation resulting in many nucleolar fragments. Each nucleolar fragment settles inside one of the nuclear pouches and is extruded from there into the cytoplasm. *Raikova* [1976] observed that paranucleolar fibrillar masses of oogonial nuclei in *Acipenser ruthenus* and *A. guldenstadti* are apparently extruded to form specific proteinaceous bodies in the cytoplasm.

The other workers reported the complete absence of nucleolar extrusions in the fish oocyte, which were believed to be mere artifacts due to mechanical disturbance during sectioning of the material [*Wallace,* 1903; *Wheeler,* 1924; *Nath and Nangia,* 1931; *Nath* et al., 1944; *Raven,* 1961; *Lal,* 1963]. Results of recent studies as described above do not support this suggestion.

According to many early authors, the nucleolar extrusions play a direct role in the formation of protein yolk or intervesicular yolk [*Subramaniam and Aiyar,* 1935; *Narain,* 1930; *Hisaoka and Firlit,* 1962; *Malone and Hisaoka,* 1963; *Malhotra,* 1963; *Shahi* et al., 1979; *Shrestha,* 1980]. But such a direct role in vitellogenesis has been denied by other workers [*Eggert,* 1929; *Narain,* 1930; *Singh and Boyle,* 1938; *Chaudhry,* 1951; *Guraya,* 1965; *Bose and Bose,* 1964; *Rastogi,* 1968a; *Guraya* et al., 1975, 1977]. *Kapoor* [1977] believes that the extruded nucleoli disappear near the periphery of the growing oocyte in *P. ticto* and give rise to intravesicular yolk in the peripheral vacuoles (or cortical alveoli). Their direct involvement in the formation of cortical alveoli is not supported by recent electron microscope studies as the cortical alveoli in the fish are formed by the Golgi complex as will be discussed in Chapter IV. According to *Chowdhery* et al. [1979], after nucleolar extrusions, the protein yolk precursors appear at the periphery of the oocyte adjacent to the follicular epithelium in *C. fasciatus.*

In the light of recent research findings, however, it can be suggested that the most important function of nucleolar extrusions in the growing oocytes of fish is to contribute precursors for ribosomes to the ooplasm as already discussed [see also *MacGregor,* 1972; *Davidson,* 1976]. The extrusion of nucleoli as such may, in some way, be related to the immediate accumulation of abundant ribosomes in the rapidly growing, previtellogenic oocytes of teleosts. According to *Buslé* [1980] the main cytological changes (increase in size of cell, nucleus, nucleolus, and increase in number of nucleoli, ribonucleoprotein particles, nuage and mitochondria) demonstrated with electron microscopy in the previtellogenic oocytes of golden grey mullet *(Mugil auratus)* provide evidence for important synthetic processes in an early preparatory phase of oocyte development. The precise nature and regulation of such synthetic processes must be understood at the molecular level with recombinant DNA techniques.

III. Cytoplasmic Components

Corresponding to various morphological, histochemical and biochemical changes of the nucleus as discussed in chapter II, there also occur conspicuous alterations in the cytoplasm of growing oocyte in fish ovary. The various structures of the follicle during previtellogenesis are directed towards the synthesis of cytoplasm and its components as evident from the proliferation, accumulation and differentiation of various cytoplasmic structures namely, yolk nucleus substance (ultrastructural elements of endoplasmic reticulum and ribosomes), free ribosomes (RNA), mitochondria, Golgi bodies, lipid bodies, etc. Some of these ooplasmic components form a paranuclear complex in the young oocytes, which is called the Balbiani's vitelline body [*Guraya,* 1979b]. During previtellogenesis (or the primary growth phase) the oocyte volume increases a few hundredfold, the nucleocytoplasmic ratio decreases. Most studies have indicated that the primary growth phase oocurs independent of pituitary influence [*Khoo,* 1979]. After hypophysectomy, oocytes do not grow beyond a 'critical stage' [*Fostier* et al., 1983; *Ng and Idler,* 1983]. Although the absolute critical size is species-dependent, it is constant within a species [*Barr,* 1968a]. By the time this size is reached, the oocyte lies within its definite follicle; it is immediately surrounded by a zona pellucida, a single layer of squamous follicle cells, by a layer of thecal cells, and most externally by a layer of surface epithelial cells as will be discussed in Chapter V later on.

A. Balbiani's Vitelline Body

Very divergent views were expressed about the origin, structure, nature and function of Balbiani's vitelline body [*Raven,* 1961]. Recent electron microscope, histochemical and autoradiographic studies have revealed the details of the origin, differentiation, structure, nature and function of Balbiani's vitelline body in the oocytes of different fish species, which will be described here. Balbiani's vitelline body in the fish oocyte usually forms a paranuclear complex of organelles and inclusions of diverse morphology and chemical nature, which include basophilic yolk nucleus, mitochondria, lipid bodies, Golgi bodies, etc. [*Guraya,* 1979b]. The yolk nucleus in the young oocytes of the dogfish *(Scoliodon sorrakowah)* gradually develops adjacent to the nuclear envelope and finally forms a large juxtanuclear mass of subspherical shape (fig. 29) [*Guraya,* 1979b]. It shows a central spherical or hemispherical zone of a more dense, organized nature, and a peripheral zone of loose consistency; both zones may be separated by a transitional zone of clear cytoplasm (fig. 29c). Sometimes, several dense, organized bodies, which, in their texture and morphology, resemble the central zone of the yolk nucleus, are seen in the outer intensely basophilic substance having a granular appearance (fig. 29d). Mitochondria and lipid bodies also associate with the yolk nucleus.

Relatively, Balbiani's vitelline body has been more extensively studied in the oocytes of teleosts. A granular, basophilic substance gradually accumulates in the form of a crescent or cap adjacent to the nuclear envelope (fig. 30). Finally, it forms a large juxtanuclear dense mass of hemispherical shape, which has been designated as the yolk nucleus (fig. 21a, b, 27, 30) by *Guraya* [1963, 1965 – see also *Kraft and Peters,* 1963; *Kumari and Padmanabhan,* 1976; *Kapoor,* 1977; *Sobhana and Nair,* 1977; *Shahi* et al., 1979; *Chowdhery* et al., 1979; *Takahashi,* 1981].

The development and differentiation of the yolk nucleus in the oocytes of teleosts is closely accompanied by a proliferation of mitochondria and phospholipid bodies, for which the former constitutes the substrate. Thus the Balbiani's vitelline body of the teleost oocyte is composed of a yolk nucleus, mitochondria and phospholipid bodies (fig. 21a–c, 27) as also reported for the dogfish oocyte (fig. 29) [*Guraya,* 1979b]. *Nayyar* [1964] observed that the Balbiani's vitelline body of the teleost *(Heteropneustes fossilis)* consists of lipids and mitochondria and reported that its function was to initiate the synthesis of lipids. Balbiani's vitelline body in the oocytes of *Merluccius* originates as an accumulation of mitochondria, Golgi bodies, lipids, RNA, protein and glycogen [*Ramadan* et al., 1979b]. *Nayyar* [1964] and *Ramadan* et al. [1979b] overlooked the yolk nucleus proper, which constitutes the substrate for other components such as the mitochondria and phospholipid bodies [*Guraya,* 1963, 1965].

Balbiani's vitelline body in the oocytes of *Channa* may show one or two dense bodies composed of RNA and protein [*Guraya,* 1963], which may be the extruded nucleoli [*Kraft and Peters,* 1963; *Guraya,* 1965]. *Gopal Dutt* [1964] also reported a similar dense structure in the yolk

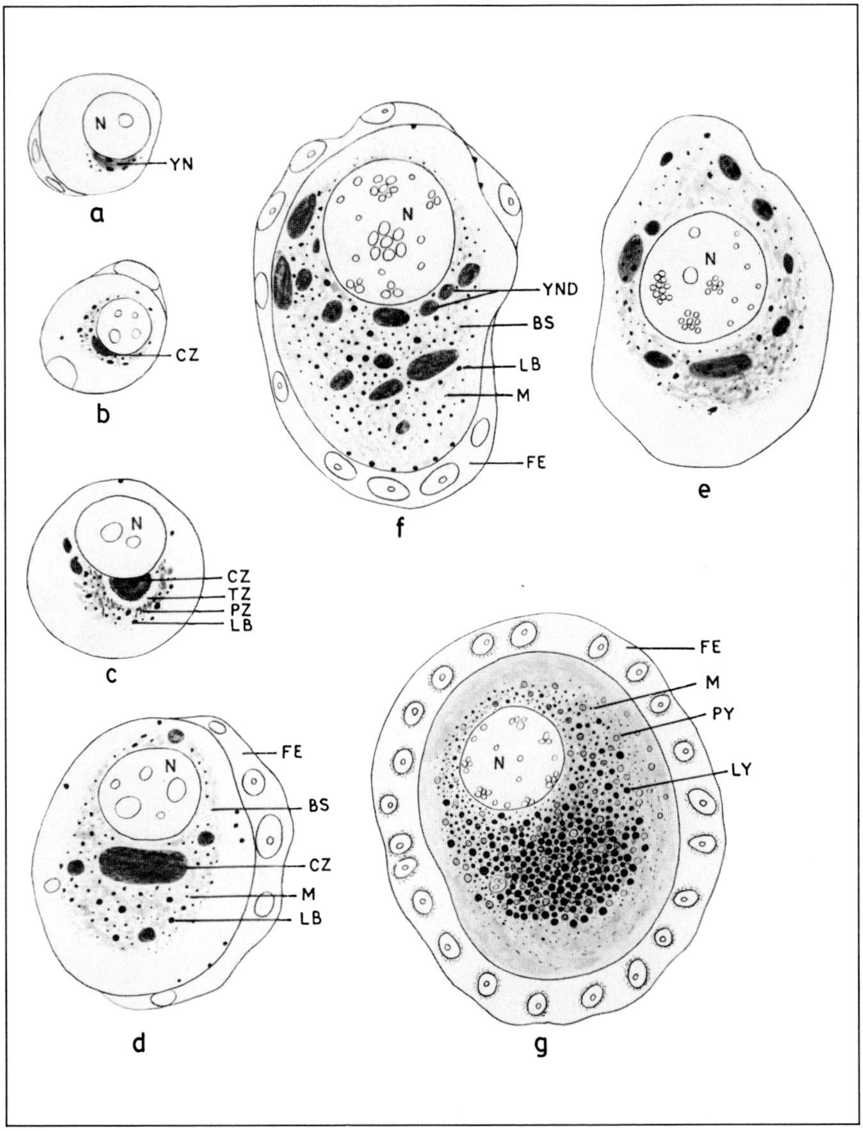

Fig. 29a–g. The origin, development, morphology, and behaviour of the yolk nucleus and other cell components of the Balbiani's vitelline body in successive stages (*a–g*) of oocyte growth in the dogfish *(S. sorrakowah).* BS = basophilic substance; CZ = central zone of yolk nucleus; M = mitochondria; N = nucleus of oocyte showing nucleoli; PY = protein yolk, PZ = peripheral zone of yolk nucleus; TZ = transitional zone of yolk nucleus; YN = yolk nucleus; YND = yolk nucleus derivatives [from *Guraya,* 1979b].

nucleus of *Anabas scandens* oocytes, which appears to arise adjacent to the nuclear envelope and is reactive for RNA. He has not paid much attention to the mitochondria and lipid bodies closely associated with the yolk nucleus to constitute the Balbiani's vitelline body.

The basophilic yolk nucleus of Balbiani's vitelline body gives positive histochemical reactions for RNA, protein and some lipoprotein (fig. 30) [*Guraya,* 1963, 1965, 1979b; *Abrahim* et al., 1966]. Mitochondria consist of lipoproteins. The lipid bodies are composed of phospholipids. *Livni* [1971] described the following enzymes to be present in the Balbiani's vitelline body of three species of teleost: glucose-6-phosphate dehydrogenase, α-glycerophosphate dehydrogenase, and succinate dehydrogenase.

Electron microscope studies have revealed that during the primary growth phase, most teleost oocytes accumulate extensive aggregations of basophilic and electron-dense material in the perinuclear cytoplasm [*M. Yamamoto,* 1964; *Droller and Roth,* 1966; *Zahnd and Porte,* 1966; *Anderson,* 1968; *Ulrich,* 1969; *Wegmann and Götting,* 1971; *Beams and Kessel,* 1973], which is identical to the basophilic, RNA-containing yolk nucleus of light microscopy [*Guraya,* 1963, 1965, 1979b]. *Beams and Kessel* [1973] using light and electron microscopy followed the behaviour of the yolk nucleus in the Balbiani's vitelline body from its origin in the youngest oocyte to its disappearance at the beginning of vitellogenesis in the trout. It first develops adjacent to the juxtanuclear ooplasm, where it shows the form of a duplex body consisting of a nonbasophilic component (an idiosome) and a highly basophilic portion (the pallial substance) as already described in the yolk nucleus of dogfish oocytes [*Guraya,* 1979b]. The yolk nucleus of trout oocyte forms a dense, finely granular substance apparently unrelated to the Golgi apparatus or mitochondria. *M. Yamamoto* [1964] observed that the yolk nucleus of the Balbiani's vitelline body in the oocytes of *Oryzias* consists of a network of small granules which, according to him, are probably RNP and of nuclear origin. This is supported by the fact that *Ulrich* [1969], in the teleost *Panio,* has observed the migration of nuclear ribosomes through the pores of the nuclear envelope into the perinuclear ooplasm where they accumulate in small groups and often develop a close association with mitochondria. *Wegmann and Götting* [1971] also described a network of granules associated with development of the Balbiani's vitelline body in the teleost *Xiphophorus.* According to *Riehl* [1976a, 1978a], the Balbiani's vitelline body (yolk nucleus) in young oocytes of the freshwater teleosts *Noemacheilus barbatulus* and *Phoxinus phoxinus* is composed of a nucleolar substance and mitochondria and in *Noemacheilus*

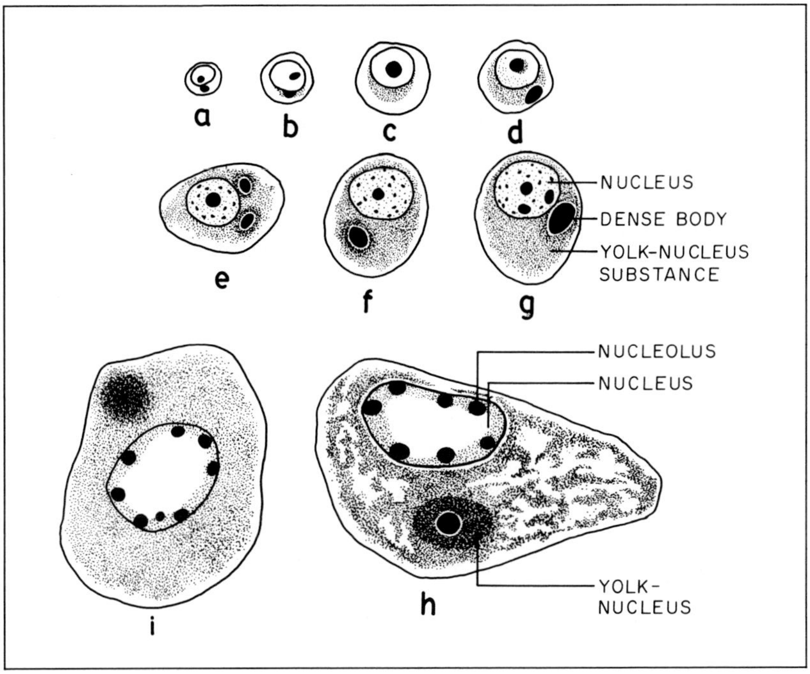

Fig. 30. Oocytes of fish, drawn from material fixed in Zenker's fluid and stained with methyl green pyronin. *a* Very young oocyte of *Channa,* showing dense body. *b* Very young oocyte of *Channa,* showing basophilic yolk nucleus substance and a dense body adjacent to the nuclear envelope. *c* Very young oocytes of *Heteropneustes* showing yolk nucleus substance in the perinuclear cytoplasm. *d–g* Young oocytes of *Channa,* showing the growth of a yolk nucleus and dense body. *h* Growing oocyte of *Channa* showing the proliferation of basophilic masses from the yolk nucleus. *i* Large previtellogenic oocyte of *Channa,* showing a homogeneous yolk nucleus and uniformly distributed basophilic substance [from *Guraya,* 1963].

partly of dictyosomes (fig. 31a, b). The osmiophilic nucleolar material, which is identical to the RNA-containing yolk nucleus substance of light microscopy [*Guraya,* 1963, 1965, 1979b], leaves the nucleus through the pores of the nuclear envelope and aggregates in the juxtanuclear ooplasm where it is surrounded by mitochondria (fig. 31b; see also Chapter II). The yolk nucleus substance contains mainly RNA as further confirmed by marking with [^{3}H]-uridine [*Riehl,* 1978a].

Clérot [1976] studied the ultrastructural features of the juxtanuclear complex of organelles (or Balbiani's vitelline body) in young oocytes of several species of cyprinid fish, which consist of intermitochondrial 'cement', annulate lamellae, mitochondria, and so on. Annulate lamellae develop as a part of the intermitochondrial cement which corresponds to the nucleolar or yolk nucleus substance of other investigators and also originates from the nucleus. Intermitochondrial cement appears to be involved in active biogenesis of mitochondria. *Toury* et al. [1977] observed that isolated intermitochondrial cement (of cyprinid teleost *Tinca tinca*) from germ cells contains 28S and 5S RNA, proteins, cytochromes and lipids. RNA and protein are the major constituents of the cement (or yolk nucleus) proper, thus supporting the cytochemical and autoradiographic data. Lipids are shown to be contaminants from the inner mitochondrial membranes. Analysis of the isolated cement by polyacrylamide gel electrophoresis resolves 20 protein bands, most of which show the same mobilities as mitochondrial proteins. Among the mitochondrial enzymatic activities tested in the isolated cement, only cytochrome oxidase and NADH-dehydrogenase are detected. The majority of the proteins present in cement are apparently destined to be incorporated into the mitochondria being formed in the clusters. *Toury* et al. [1979] have further extended and confirmed these various observations in regard to perinuclear ooplasmic structures consisting of mitochondria and intermitochondrial cement in oogonia and oocytes of *T. tinca.* The intermitochondrial cement is connected with nucleoplasmic nucleoprotein granules by means of material which leaves the nucleus through the nuclear pores (see also Chapter II), thus supporting the observations of *Riehl* [1976a, 1978a]. During the very active mitochondrial multiplication, protein synthesis takes place within the cement as demonstrated with autoradiography; the newly synthesized polypeptides being subsequently incorporated into the miochondria which increase in number. The perinuclear bodies similar to the intermitochondrial cement described in the germ cells of *Tinca* are also seen in the oogenesis of *Xiphophorus helleri* [*Azevedo and Coimbra,* 1980]. But in the oocytes of *X. helleri,* the cement is not consistently labeled, perhaps because the time intervals used are not

---

Fig. 31. a Electron micrograph of portion of a very young oocyte of the fish *N. barbatulus* (L.) showing the nucleolar substance (6), mitochondria (7), dictyosome (8), and ribosomes of a developing Balbiani's vitelline body adjacent to the nuclear envelope. Nucleoplasm (1), nuclear envelope (2) and nuclear pores (3) are also seen. *b* Electron micrograph of a portion of a young oocyte of *N. barbatulus,* showing the yolk nucleus

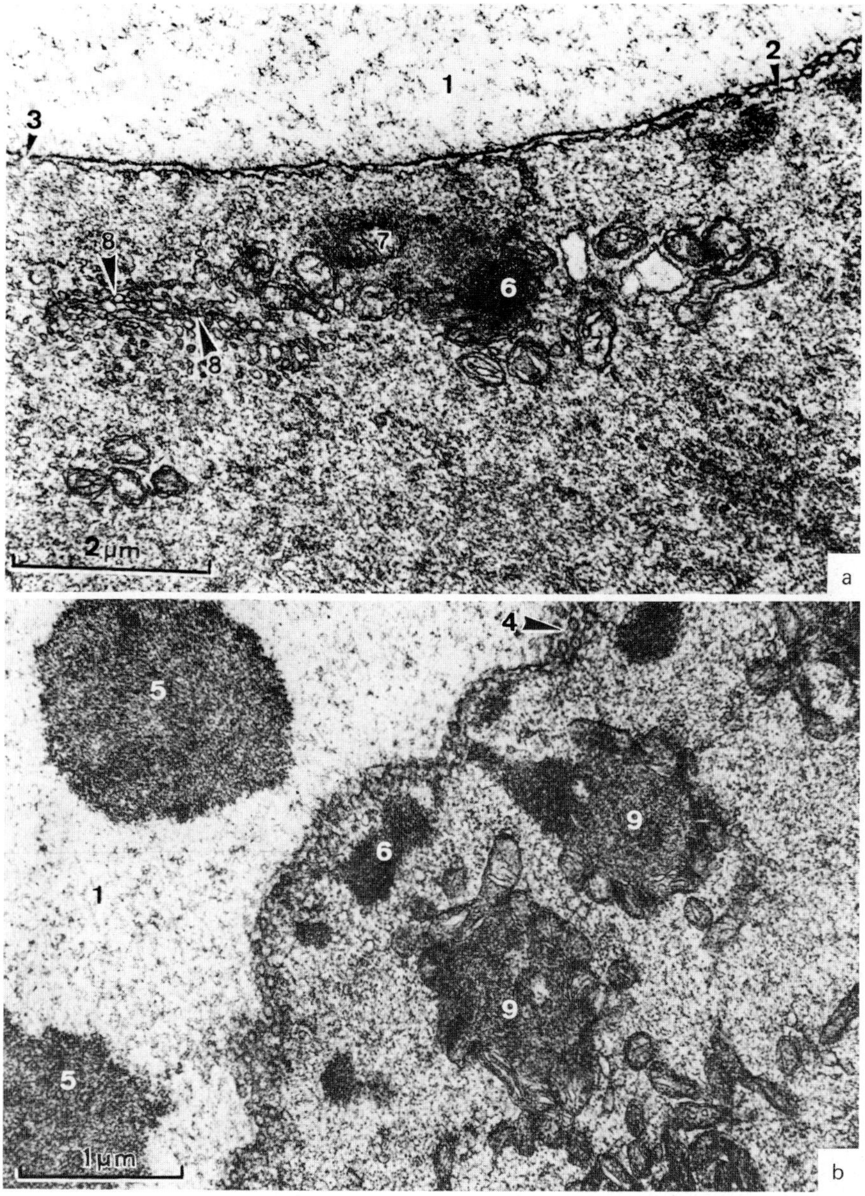

oocyte of *N. barbatulus,* showing the yolk nucleus substance (9) associated with mitochondria, which originates from the nucleolar substance (6); the latter resemble the nucleoli (5) very closely [from *Riehl,* 1976a].

long enough to allow RNA to migrate from the nucleus. All these observations suggest that intermitochondrial cement consisting of RNA and protein is involved in the synthesis of proteins for the construction of other organelles such as the mitochondria. The technique of microinjection of specific monoclonal antibody into young oocytes will be very useful to identify proteins that are involved in the regulation and proliferation of organelles in growing oocyte.

Various components of the fully developed Balbiani's vitelline body in the oocytes of the dogfish ovary are dispersed while still lying in the juxtanuclear cytoplasm (fig. 29e–g). With the oocyte growth in teleosts, the Balbiani's vitelline body or juxtanuclear complex of organelles separates from the nuclear envelope and persists as a spherical structure (fig. 2a, 21b, c, 30h, i) [*Chaudhry,* 1952; *Guraya,* 1963, 1965; *Kraft and Peters,* 1963; *Gopal Dutt,* 1964; *Nayyar,* 1964; *Wegmann and Götting,* 1971; *Guraya* et al., 1975, 1977]. Meanwhile, the yolk nucleus of the Balbiani's vitelline body produces from its peripheral portions a basophilic substance that is rich in RNA (fig. 30h) [*Guraya,* 1963, 1965]. This substance shifts into the outer ooplasm as irregular masses. The mitochondria also multiply in number and develop morphological association with the masses of basophilic material during previtellogenesis (fig. 21b, c). Ultimately, the Balbiani's vitelline body moves to the cortical ooplasm (fig. 2a, 30i) where it finally disintegrates [*Kraft and Peters,* 1963; *Chowdhery,* 1979; *Shahi* et al., 1979]. The factors regulating the development, differentiation, peripheral migration and eventual disintegration of Balbiani's vitelline body in growing oocyte of teleost must be determined in future studies. However, the protein yolk bodies in the oocytes of dogfish and teleosts make their first appearance in zones of the ooplasm, which are rich in RNA-containing yolk nucleus substance and mitochondria derived from the Balbiani's vitelline body (fig. 29g) [*Guraya,* 1965, 1979b; *Ramadan* et al., 1979b]. The stage I oocytes of *Neomacheilus* and *Phoxinus* contain up to 40 yolk nuclei which can be only demonstrated by electron microscopy [*Riehl,* 1976a, 1978a]. The yolk nuclei of both species dissolve in early stage III oocytes. Their components are distributed in the ooplasm. *Beams and Kessel* [1973] stated that the exact role of yolk nucleus in oogenesis is still not clear but that it probably forms an essential precursor substance (perhaps RNA) needed for oocyte growth and vitellogenesis, which corresponds to the RNA-rich yolk nucleus substance of *Guraya* [1963, 1965]. *Riehl* [1976a] believes that the Balbiani's vitelline body (yolk nucleus) has no relationship to the formation of yolk in the oocyte of *N. barbatulus.*

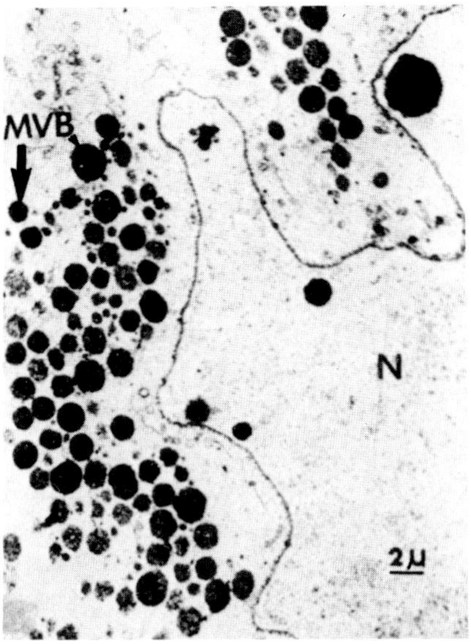

Fig. 32. Electron micrograph of portions of oocyte from the ovary of juvenile rainbow trout treated with salmon gonadotrophin showing large number of multivesicular bodies (MVB) around the oocyte nucleus (N) [from *Upadhyay* et al., 1978].

From the correlation of recent cytological (including ultrastructural), histochemical and biochemical (including autoradiographic) data, it can be concluded that the Balbiani's vitelline body in oocytes of different fish species shows the accumulation of different organelles and inclusions of diverse morphology and nature [see also *Guraya,* 1979b]. Therefore, it shows a great variability in development, organization, and composition among different species of fish. Depending upon the fish species, Balbiani's vitelline body shows different combinations of various organelles and inclusions such as a yolk nucleus (or ultrastructural RNP particles or cement and element of endoplasmic reticulum), mitochondria, Golgi bodies, annulate lamellae, multivesicular bodies (fig. 32) [*Upadhyay,* 1977; *Upadhyay* et al., 1978], lipid inclusions, and so on, which aggregate to a variable degree in the perinuclear ooplasm. Of these organelles and inclusions, the yolk nucleus proper consisting of RNA and protein forms the most conspicuous feature of teleost oocyte, and generally persists as an organized structure throughout various stages of previtellogenesis; meanwhile it contributes its substance to the outer ooplasm. Balbiani's vitelline body as a whole acts as the centre of metabolic activities related to the formation, multiplication and accumulation of other organelles and inclusions which are finally distributed in the outer ooplasm, before the deposition of yolk. These organelles and materials are needed within the oocyte prior to yolk formation. Another enigmatic body ocassionally seen in primary growth stage oocytes of certain teleosts is a large island of highly staining cytoplasm [*Wallace and Selman,* 1980]; electron microscope study of this region does not demonstrate components of any kind. Nevertheless, it also disperses in a manner similar to that observed for the Balbiani's vitelline body. In future studies, the Balbiani's vitelline bodies or yolk nuclei of fish oocyte must be isolated for the chemical characterization of their components which may consist of various RNA species and proteins with structural and enzymic functions. A thorough understanding of the nature of such chemical components is needed to define the function(s) of Balbiani's vitelline bodies in the biology of oocyte growth.

B. Mitochondria

Light microscope studies revealed that the mitochondria in the growing oocytes of fishes appear as discrete granules and filaments (fig. 21, 22) [*Raven,* 1961; *Guraya,* 1965, 1978, 1979b; *Chopra,* 1958a, b, 1960; *Malone and Hisaoka,* 1963]. Histochemical studies demonstrated their lipoprotein composition. *Ramadan* et al. [1979a] also observed that mitochondria in early oocytes of *Merluccius merluccius* appear as fine granules, with a slight perinuclear concentration. As oocyte grows, they show linear alignment. Later, the cytoplasm appears as a mesh-like structure with the mitochondria attached to its interlacing threads which may be constituted by the yolk nucleus substance (or ultrastructural elements of ER and ribosomes). *Guraya* [1963, 1965] also demonstrated a close morphological relationship between the yolk nucleus substance and mitochondria of growing previtellogenic oocytes of teleosts *(Channa marulius* and *Heteropneustes fossilis)* (fig. 21b, c, 22a, b) suggesting a close functional relationship between them. In the advanced stages of oocyte growth, the mitochondria are distributed among the large yolk droplets occupying the spaces between them.

Various electron microscope studies have shown that mitochondria undergo changes in their distribution, amount, shape, size and internal structure during oocyte growth in teleosts, which are indicative of metabolic alterations during oocyte growth. In the oocytes at the beginning of cytoplasmic growth, the mitochondria are, as a rule, concentrated around the nucleus or germinal vesicle and as the growth proceeds, are distributed all over the cytoplasm. In the oocytes of *Brachydanio rerio,* the small, numerous mitochondria are localized in the perinuclear zone but the large ones are distributed in the peripheral regions [*Ozernyuk* et al., 1980], suggesting the presence of some regional metabolic differences. *Toury* et al. [1979] observed that in oogonia of *Tinca tinca,* the number of mitochondria is small. It increases enormously during the growth of oocyte. During this period, the mitochondria are clustered around spheres of an electron-dense opaque material, the intermitochondrial cement. These clusters are placed close to the nucleus. The mitochondria in the oocytes of *Salmo gairdnerii* R. vary in their size, shape and internal structure (fig. 3) [*Upadhyay,* 1977]. A few mitochondria are sometimes clustered around a dense granular substance which resembles the material extruded from the nucleus. Mitochondria of early oocytes in guppy *(Lebistes reticulatus)* are large and contain both tubular and shelf-like cristae [*Droller and Roth,* 1966]. The latter lie near the periphery of the mitochondria, frequently parallel to their envelope. The tubular cristae extend into the matrix in a direction perpendicular to the long axis of mitochondrion. With the oocyte growth, the mitochondria increase in number and meanwhile, some of these become oblong in shape and thinner (fig. 6, 33). In addition, tubular cristae no longer characterize any of the mitochondria in large previtellogenic oocytes. With the start of lipid yolk deposition, the mitochondria are predominantly of the thinner diameter and appear to be uniformly distributed throughout the oocyte cytoplasm (fig. 33). *Caloianu-Iordachel* [1978] has discussed the changes in the mitochondria, Golgi complex, endoplasmic reticulum (ER) and other organelles during oocyte growth in Acipenserid fishes *(Acipenser guldenstadti* and *A. stellatus).*

Ozernyuk and Kotomin [1976] observed that the growth of loach oocytes is closely accompanied by an increase in mitochondrial density. Corresponding to the increase in density, there occurs an acceleration of [^{14}C]-valine incorporation into mitochondrial proteins. The increase in mitochondria density during oogenesis is probably due to an increase in the amount of membrane material per unit of mitochondria weight. The measurement of specific activity of cytochrome oxidase with parallel determination of protein/phospholipid ratio in mitochondria of loach *(Misgurnus fossilis)* oocytes has been used in analysis of mitochondrial membrane growth [*Ozernyuk,* 1981]. Specific activity of cytochrome oxidase in mitochondria in the course of oogenesis from the stage of small growth of oocytes to the stage of the beginning of large growth is increased 57%. Apparently this effect is related to increased enzyme amount per milligram of mitochondrial protein, i.e., the increase of the protein/phospholipid ratio in mitochondria takes place at these stages. The rate of respiration and concentration of mitochondrial protein are increased during early stages of oogenesis in *Brachydanio rerio* and decreased during later stages [*Ozernyuk* et al., 1980]. Increase of the protein mass during oocyte growth is conditioned by two parallel processes, growth and division of mitochondria in oocytes. *Ozernyuk* [1981] has observed that the specific activity of cytochrome oxidase in the fraction of heavy mitochondria obtained from loach oocytes is twice that in the fraction of light mitochondria. The cytoplasm of oogonia and early diplotene oocytes in the ovaries of silver carp is characterized by lesser electron density compared with the oocytes during cytoplasmic growth [*Emel'Yanova,* 1979].

C. Golgi Complex

Fish oocytes treated with classical techniques of cytology showed Golgi complex in the form of granules, spheres, crescents, etc. [*Raven,* 1961]. The Golgi apparatus in the oocytes of *Merluccius merluccius* also occurs in the form of granules of different size and spheres in addition to some crescents and rings showing argentophil cortex and argentophobe medulla [*Ramadan* et al., 1979a]. In the advanced stages of oocyte growth, the Golgi bodies lie

---

Fig. 33. Portion of guppy oocyte during proteid yolk formation showing numerous dark, crenate inclusions (ci) and lighter, spherical inclusions (py), both membrane-limited, among the frequent Golgi bodies (G), mitochondria (m), and ER vesicles (er) of cortical cytoplasm. Asterisks show ER vesicles in association with the large spherical inclusions (py). Coated pits and vesicles (cp) appear between the microvilli (mv) of the zona pellucida (zr). Vitelline membrane (vm) is deposited between the microvilli and the surface of the overlying follicle cells (fe). The inset shows several features of intraoocyte yolk formation. Smooth vesicles of ER (er) in the cortical cytoplasm contain flocculent material. Two Golgi complexes (G) are filled with a material resembling that in 70-μm diameter vesicles and crenate inclusions (ci). Aggregates of several independent vesicles (iva) can also be seen. × 12,500; inset × 14,500 [from *Droller and Roth,* 1966].

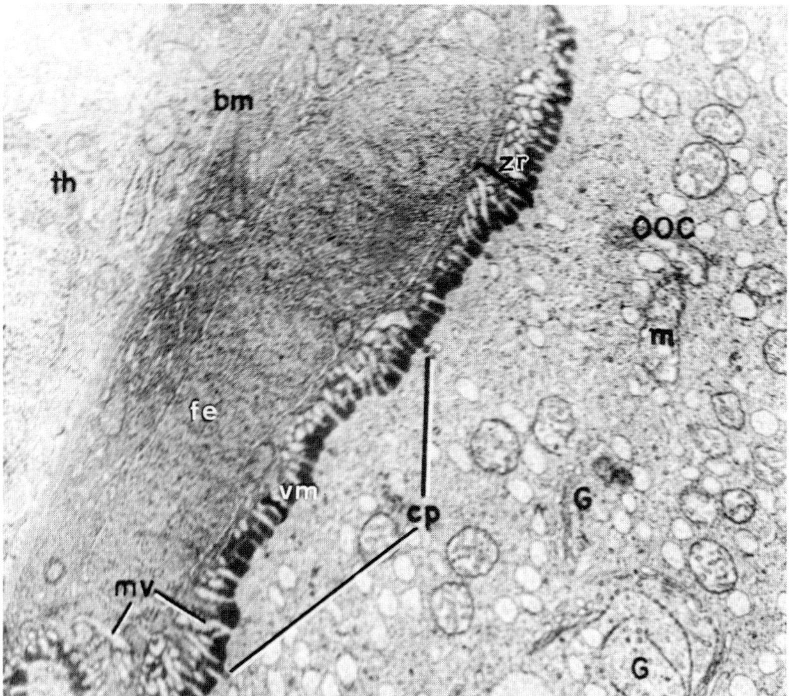
bm
th
zr
ooc
m
fe
vm
cp
G
mv
G
* *
er
ci
ci
er
er
pv
G
G
iva
er

between the cortical alveoli of peripheral ooplasm where they play some significant roles in the elaboration/secretion of materials needed for the formation of cortical alveoli, zona pellucida, lysosomes, endogenous yolk (in some species), etc. Such functions of Golgi apparatus still need to be resolved precisely at the subcellular and molecular level in the growing oocyte of fish. The Golgi complex is rarely seen in oogonia and oocytes of silver carp [*Emel'Yanova,* 1979]. *Droller and Roth* [1966] using electron microscopy observed that a few Golgi systems, which occupy the oocyte cytoplasm from the time when the egg is in the nest until it is sequestered in a separate follicle, are characterized by randomly arranged cisternae and several small vesicles (fig. 5). The Golgi complex of the cortical cytoplasm bear no close topographical association with other components of the oocyte (fig. 33, 34). However, several cisternae and vesicles of these complexes show an electron-opaque substance. With the initiation of intraoocyte yolk formation, there are seen many Golgi complexes in the oocyte of cytoplasm. Most of them contain an electron-opaque material of a coarse, granular consistency. The Golgi apparatus in the oocytes of *Salmon* consists of three to four stacks of flattened curved cisternae and a few small vesicles (fig. 3) [*Upadhyay,* 1977]. Further studies are needed to reveal the details of origin, multiplication, differentiation and functions of Golgi complex in the developing oocytes of different fish species.

D. Endoplasmic Reticulum

In its development and distribution, endoplasmic reticulum (ER) shows some variations in the growing oocytes of different teleosts (fig. 3, 5, 33, 34) [*Droller and Roth,* 1966; *Beams and Kessel,* 1973; *Caloianu-Iordachel,* 1978; *Bruslé,* 1980]. The functional significance of these variations in the biology of oocyte growth is not known. At first, the rough-surfaced ER of the primary oocyte in the guppy is present only in cisternal form (fig. 5) [*Droller and Roth,* 1966]. As the oocytes subsequently are segregated into separate follicles, the ER appears in the form of vesicles that are largely smooth-surfaced (fig. 33, 34). During these early stages of oocyte growth, they do not show resolvable content. Gradually, the ER vesicles in the cortical ooplasm accumulate a somewhat flocculent, loosely-packed material within their limiting membrane. In the deeper regions of the oocyte cytoplasm, vesicles that have a ribosomal coat also begin to show an electron-opaque substance. The substances in both types of vesicles are mor-

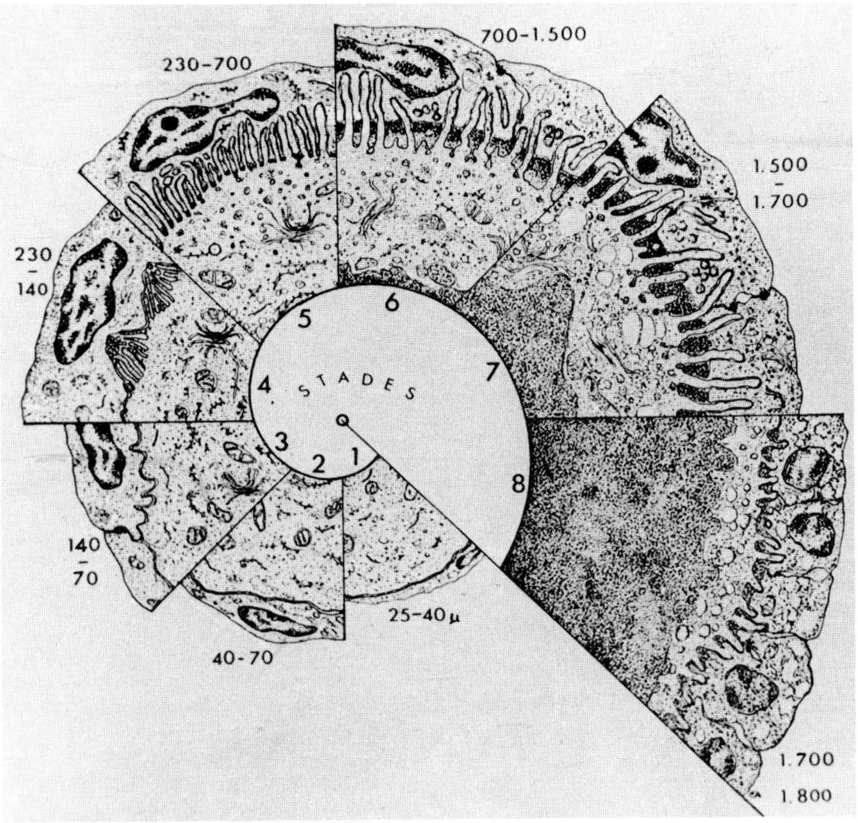

Fig. 34. Diagrammatic representation of morphology and distribution of ooplasmic organelles, and development and morphology of various components of zona pellucida and follicular epithelium during successive stages (stades) of oogenesis in *Xiphophorus* [from *Azevedo,* 1974].

phologically indistinguishable. They are quite distinct though from the densely granular content of the Golgi complexes. Electron microscope studies revealed that, at least, in early stages of vitellogenesis, there occurs an extensive development of granular endoplasmic reticulum in the ooplasm. The ER in the oocytes of *Salmon* is vesicular or tubular in form and is generally agranular (fig. 3, 35) [*Upadhyay,* 1977]. In the oocytes of silver carp, ER is also represented by vesicles, and less often, by cisternae of mainly agranular type [*Emel'Yanova,* 1979]. ER and ribosomes are identi-

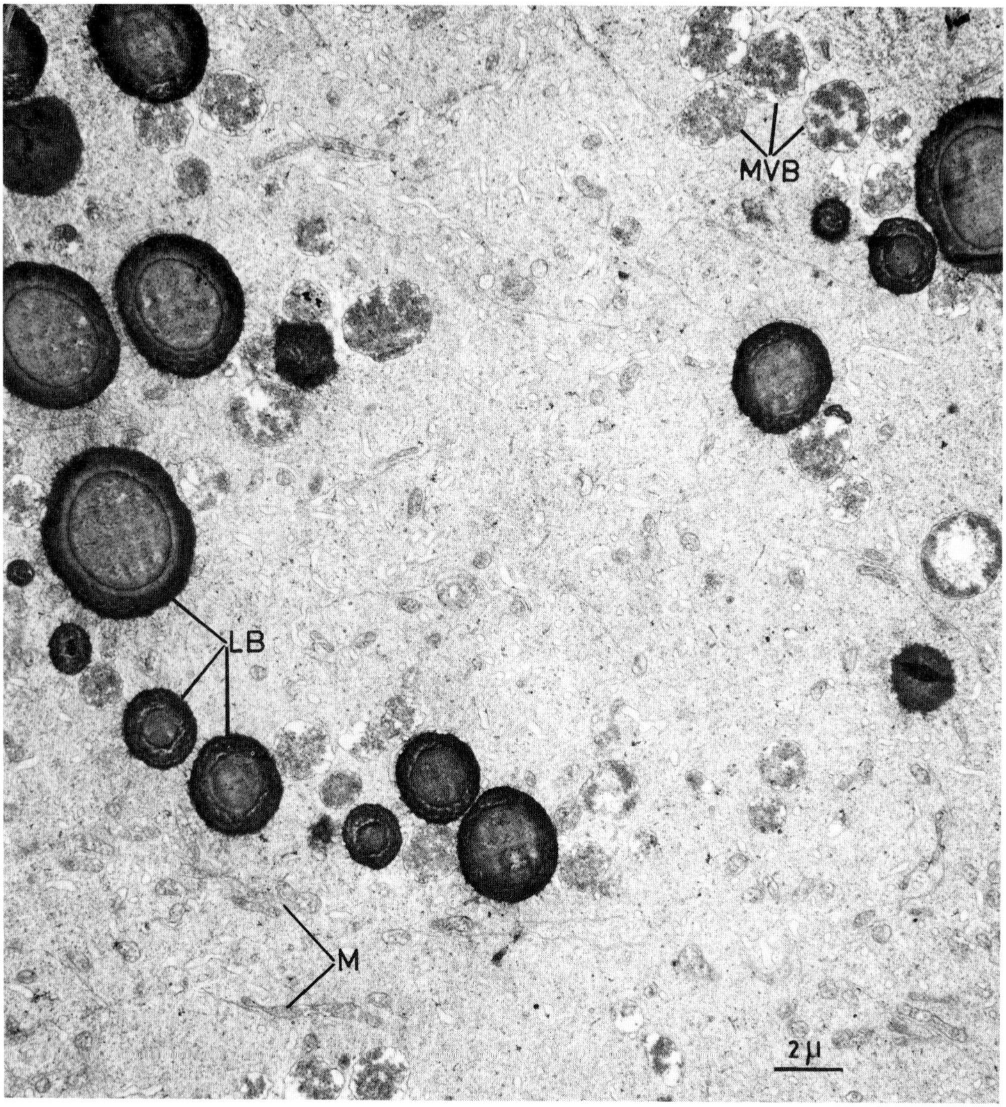

Fig. 35. Electron micrograph of portion of oocyte from the ovary of juvenile rainbow trout, showing lipid bodies (LB), having concentric zonation, multivesicular bodies (MVB), mitochondria (M) of variable size and elements of endoplasmic reticulum in the ooplasm. Some lipid bodies lie in the vicinity of multivesicular bodies [courtesy of Dr. *S.N. Upadhyay*].

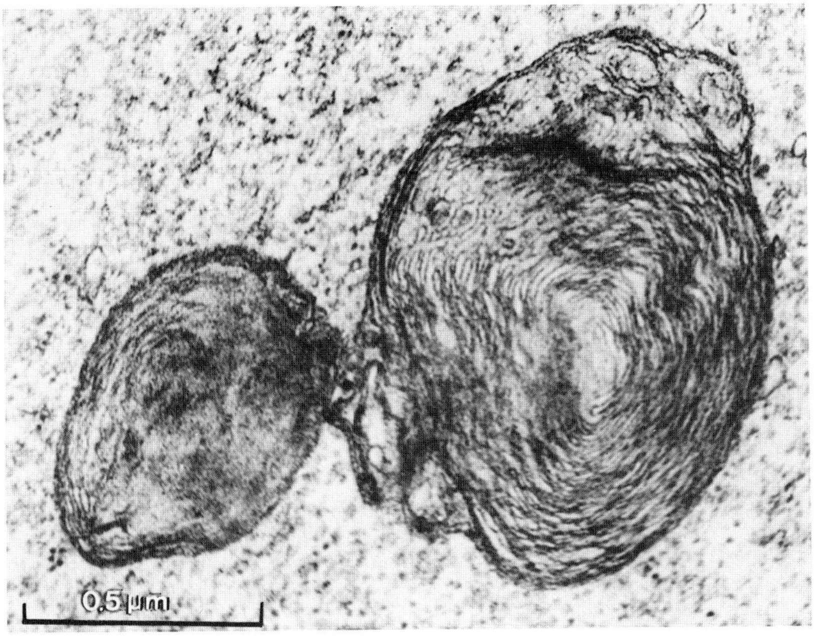

Fig. 36. Two typical lamellar bodies of oocytes in *Noemacheilus.* On the right side the inner membrane is clearly seen while on the left side, the membrane structure has been lost to a considerable degree [from *Riehl,* 1977a].

cal to the yolk nucleus substance of light microscopy [*Guraya,* 1979b]. The significance of changing pattern of ER during oocyte growth must be determined in relation to its developmental processes.

E. Lamellar Bodies

Riehl [1977a] has described lamellar bodies in the oocytes of *Noemacheilus barbatulus* (fig. 36) which are only found during the oocyte stage I. The lamellar bodies shift from the vicinity of the nucleus to the periphery of the oocyte. They appear to be of great importance during the formation of yolk. Ultrastructural lamellar bodies may be identical to the phospholipid bodies of histochemical preparations [*Guraya,* 1965]. But this suggestion needs to be confirmed with correlative electron microscope and autoradiographic studies.

F. Multivesicular Bodies

Upadhyay et al. [1978] have observed that small oocytes (175–233 μm) of juvenile rainbow trout (*Salmo gairdneri* R.) are characterized by the presence of multivesicular bodies at the perinuclear cytoplasm (fig. 32). But in the medium (285–350 μm) and large (400–495 μm) oocytes, multivesicular bodies are seen sparsely distributed throughout the ooplasm (fig. 37) [see also *Upadhyay,* 1977]. Such an extensive development of multivesicular bodies has not been reported in previtellogenic oocytes of other teleosts. The physiological meaning of multivesicular bodies in the juvenile rainbow trout should be determined.

G. Lipid Bodies

Previtellogenic oocytes of fish ovary show sudanophilic lipid bodies of variable size, and shape (fig. 21, 22, 27) which consist of phospholipids [*Chopra,* 1958a, b, 1960; *Guraya,* 1963, 1965; *Shackley and King,* 1977]. *Riehl* [1977c] observed acidic lipids in stage I oocytes of *Noemacheilus* and *Gobio. Ramadan* et al. [1979b] also found that cytoplasm of all stages of oocyte growth in *Merluccius* contains lipid bodies of variable size, which increase in number with the size of growing oocyte. *Malone and Hisaoka* [1963] did not make any reference to such lipid bodies during the histochemical studies on the oocytes of zebra fish. *Droller and Roth* [1966] using electron microscopy observed large sudanophilic lipid droplets in the oocyte cytoplasm quite early during oogenesis in *Lebistes reticulatus. Upadhyay* et al. [1978] demonstrated lipid bodies in the medium-sized (285–350 μm) and large (400–495 μm) oocytes of the juvenile rainbow trout, which are found scattered amongst the multivesicular bodies (fig. 35) [see also *Upadhyay,* 1977]. In the large oocytes, the lipid bodies are dispersed throughout the cytoplasm and show concentric zonation (fig. 35). The concentric zonation may be an ultrastructural manifestation of either various lipids present in different regions of the same lipid droplet, or variable physicochemical state of lipids. Electron microscope studies have also revealed that ultrastructure of lipid droplets changes during oocyte growth as the fully developed lipid droplets in large oocyte do not show zonation, and become electron lucent (fig. 38) indicating some chemical changes in lipid droplets during oocyte growth. In future studies, lipids must be isolated from different stages of oocyte growth for chemical analysis. Such

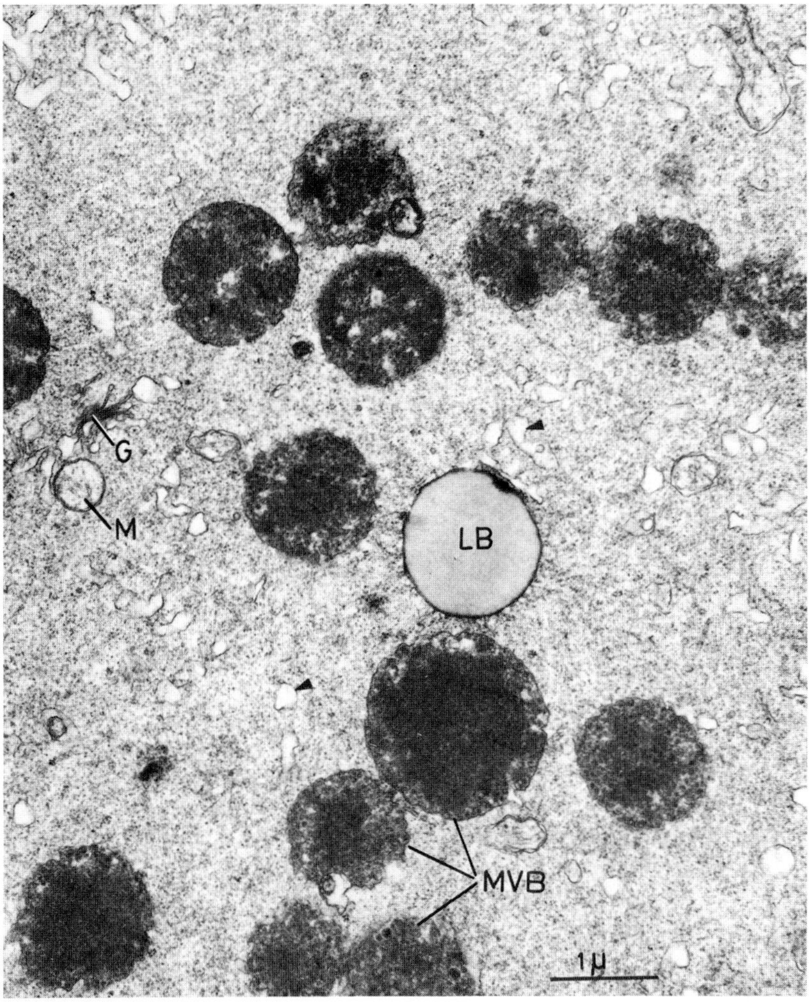

Fig. 37. Electron micrograph of portion of oocyte from the ovary of juvenile rainbow trout treated with s-GTH (0.1 μg/g body weight) showing multivesicular bodies (MVB), lipid body (LB), Golgi complex (G), mitochondria (M), elements of endoplasmic reticulum (arrowheads) and numerous ribosomes [courtesy of Dr. *S.N. Upadhyay*].

studies may help us to determine the functional significance of lipids during oocyte growth. However, both lipid bodies and multivesicular bodies of juvenile rainbow trout are believed to form constituent of endogenous yolk formed in the oocyte cytoplasm in response to salmon-gonadotrophin treatment [*Upadhyay,* 1977; *Upadhyay* et al., 1978].

According to *Guraya* [1965] lipid bodies in the growing previtellogenic oocytes of *Channa marulius* occur in the form of granules, spheres, vesicles and fenestrated plates which are generally seen adjacent to the plasma membrane of the oocyte (fig. 21, 22, 27). Some of them lie in association with spherical and tubular vacuoles which are closely related to the plasma membrane of the oocyte (fig. 22, 27). The lipid bodies and vacuoles show considerable variations in number in different oocytes and in different portions of the same oocyte. Some oocytes do not show vacuoles but continue to have some sudanophilic lipid bodies (fig. 21). The peripheral lipid bodies lie either inside or outside the vacuoles (fig. 22, 27) and resemble the sparsely distributed lipid bodies of the general cytoplasm in their histochemical reactions. They also consist of phospholipids. The sites of peripheral distribution of lipid bodies as well as their morphology and chemical composition suggest that they, in collaboration with pinocytotic vacuoles, are apparently involved in the transport of phospholipids from the follicular epithelium into the ooplasm of growing oocyte for the growth and differentiation of its various organelles [*Guraya,* 1965; see also Chapter V]. The presence of phospholipid bodies with the associations of mitochondria and yolk nucleus substance (ultrastructural ER and ribosomes) further suggests that they may be handing over their phospholipids to the latter which form the main sites of metabolic activities in the ooplasm; these phospholipids may be utilized for the synthesis of lipoprotein membranes of organelles, which multiply at the same time. Corresponding to the presence of lipid bodies in the ooplasm of growing oocyte, there also occur similar lipid bodies in its nucleus (fig. 27), as already discussed in Chapter II, indicating some close developmental and functional relationships between them during the growth of the oocyte. *Gopal Dutt and Govindan* [1967] also demonstrated lipids in the ooplasm and nucleus of all stages of oocyte growth in *Anabas.* Various lipid bodies contain triglycerides and increase in quantity without any detectable change in chemical composition during oogenesis.

A general classification of the early lipid inclusions of previtellogenic oocytes as true yolk is of questionable accuracy as these disappear in some fish species prior to, or during subsequent exogenous vitellogenesis [*Chopra,* 1958a, b, 1960; *Guraya,* 1963, 1965], but may persist in others [*Shack-

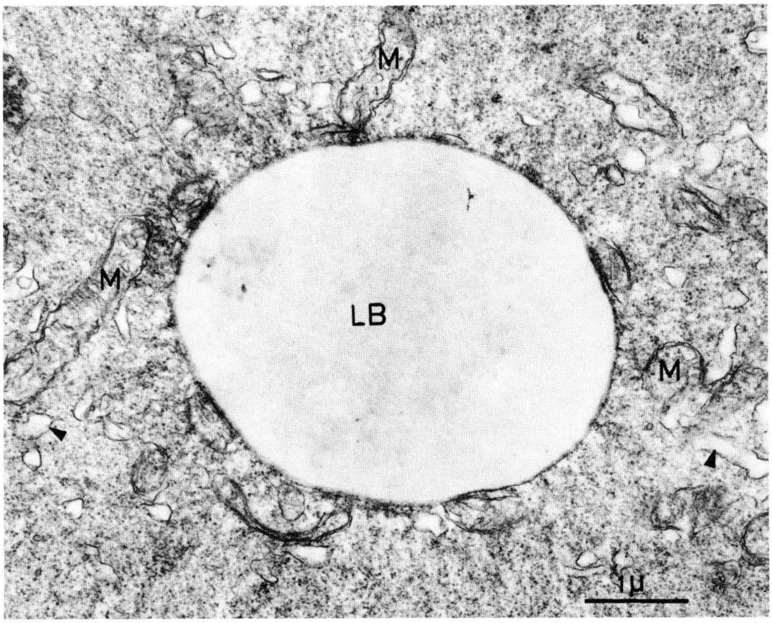

Fig. 38. Electron micrograph showing lipid body (LB), several mitochondria (M) in the vicinity of lipid body, elements of endoplasmic reticulum (arrowheads) and numerous ribosomes in the oocyte cytoplasm of the juvenile rainbow trout treated with s-GTH (0.1 µg/g body weight). The physiological significance of visible association between lipid body and mitochondria is not known [courtesy of Dr. *S.N. Upadhyay*].

ley and King, 1977]. Since such lipid inclusions consisting of either phospholipids or phospholipids and triglycerides also occur in the oocytes of other nonmammalian vertebrates [*Guraya,* 1979b] and mammals [*Guraya,* 1985], and decrease or disappear during the growth of oocyte (corresponding to the accumulation of organelles), it has been suggested that these phospholipid bodies play some important roles in the developmental processes of oocyte [*Guraya,* 1985]. However, it will be interesting to mention here that the endogenous origin of lipid bodies in the fish oocyte has been reported on the basis of morphological criteria. Detailed study of oocyte cytoplasmic lipids and direct evidence of the capacity of oocytes to synthesize them remain to be carried out with autoradiographic techniques. *Wiegand and Idler* [1982] have studied ovarian lipogenesis in vitro in the rainbow trout *(Salmo gairdneri).* Radioactive acetate incorporated into lipids has been found primarily in polar lipid fraction of ovaries in early stages of recrudescence. But it needs to be determined whether the labelled polar lipids in this study are identical to phospholipids of oocyte lipid bodies identified with histochemical techniques [*Guraya,* 1963, 1965], or to structural polar lipids, or both.

H. Ribonucleic Acid

Guraya [1963, 1965] observed that besides the basophilic yolk nucleus substance consisting of ribosomes and elements of ER, all the cytoplasm during various stages of previtellogenesis in *Channa* and *Heteropneustes* stains intensely and uniformly for diffusely distributed RNA which is not associated with any of the cell components demonstrated with light microscopy. Similar diffuse RNA has also been described in other histochemical studies of fish oocyte [*Hisaoka and Firlit,* 1963; *Sun and Xizai,* 1983], but these studies did not differentiate the basophilic yolk nucleus substance from the general diffuse basophilia. *Riehl* [1977c] observed that the oocytes of *Noemacheilus barbatulus* and *Gobio gobio* are very basophilic in stage I, showing abundant RNA in their cytoplasm. The diffusely distributed RNA may be due to the presence of various RNAs such as rRNA, tRNA, mRNA, etc., demonstrated in the oocyte of different animal species including fish [*Davidson,* 1976; *Guraya,* 1985]. These RNA species accumulate as a result of transcriptional activity of genes during oocyte growth. Stage-specific formation of these RNA species has received some attention during oogenesis of some animal groups [*Davidson,* 1976; *Guraya,* 1985], but relatively very little information along these lines is available about the fish oogenesis (see Chapter II). Therefore, it is suggested that further work using recombinant DNA technology should be carried out on fish oogenesis to determine not only the accumulation of stage-specific RNA species but also the regulation of activity of their respective genes. Abundant free ribosomes (containing rRNA) have been demonstrated in the growing oocyte of different fish species with electron microscope (fig. 38) [*Droller and Roth,* 1966; *Scharrer and Würzelmann,* 1969a, b; *Beams and Kessel,* 1973; *Riehl,* 1976a, 1978a; *Upadhyay,* 1977]. The increase in the amount of RNA and DNA is known to precede the intense accumulation of protein and other substances during oocyte growth [*Nakano,* 1969]. The stage-specific nature of proteins should be investigated in future studies. The large amount of DNA or its precursors is believed to exist not in the nucleus but in the cytoplasm as a reserve substance [*Davidson,* 1976]. The intensity of cytoplasmic and nuclear basophilia decreases as oocyte development advances [*Ramadan* et al., 1979b]. The RNA contents in the developing oocytes of *Tilapia nilotica* are believed to decrease with increasing contents of lipids, carbohydrates and proteins as determined with histochemical tests [*Sun and Xizai,* 1983]. This visible decrease of RNA may be due to the enlargement of oocyte rather than actual decrease in its amount.

Kondrat'Ev [1979] studied seasonal dynamics of distribution of RNA-rich ooplasmic structures in the developing previtellogenic oocytes of *Acipenser ruthenus.* In autumn-winter, fibrillar and fibrillar-granular substance presumably, low-M_r ooplasmic RNA were observed in oocytes 30–130 μm in diameter, which were in the early previtellogenic growth stage. This material of nuclear origin makes dense accumulations of type I and appears to be in an inactive state during the whole winter. In winter, a great part of ribosomes, fibrillar granular threads and fibrils in large previtellogenic oocytes (150–400 μm diameter formed presumably in the previous summer periods) form dense aggregations of type II in various regions of the ooplasm. In winter, part of the ribosomes is apparently eliminated from protein synthesis by making type II accumulations which are fragmented at the beginning of summer. Ribosomes forming these acucmulations are finally distributed throughout the ooplasm to initiate an active protein synthesis. These seasonal changes of RNA-rich ooplasmic structures are indicative of some metabolic alterations in oocytes, which must be determined at the molecular level in future biochemical studies. The effects of hormones on the structure, distribution and nature of ooplasmic components form a promising area for future studies as no work has been carried out along these lines.

I. Enzymes

Relatively very little effort has been made to study the enzymes of growing oocyte in fish. The production of alkaline phosphatase and acid phosphatase has been studied in the oocytes of *Silurus glanis, Perca fluviatilis* and *Acerina cernua* [*Mester and Scripcariu,* 1979]. Considerable variation is observed in these phosphatases among different teleosts [*Shaffi and Jafri,* 1975]. According to *Popov* et al. [1977] the water-soluble proteins and the enzymes lactate dehydrogenase (LDH), malate dehydrogenase (MDH), phosphatase and some esterases are species specific in Teleostei oocytes (mackerel, carp, barbel and waller – *Alosa ponticai, Cyprinus carpio, Barbus fluviatilis* and *Silurus glanis,* respectively). It is suggested that the molecular polymorphisms of LDH are less pronounced in the Teleostei than in Chondrostei fish. The oocytes of the Teleostei species show a single MDH enzyme fraction which appears to be localized in the cytosol. ADP and ATP are apparently hydrolyzed by the nonspecific phsophatases from the oocytes much faster than other naturally occuring substrates in these cells. *Sun and Xizai* [1983] have observed that activity of LDH, succinic dehydrogenase (SDH) and MDH are decreased in the growing oocyte of *T. nilotica.* However, the activity of lipase is increased. *Ozernyuk* [1981] has measured the specific activity of cytochrome oxidase with parallel determination of protein/phospholipids ratio in mitochondria of loach *Misgurnus fossilis.* Its specific activity is increased 57% during oogenesis. Apparently, this effect is related to increased enzyme amount per milligram of mitochondrial protein, i.e., the increasing of the protein/phospholipid ratio in mitochondria takes place during oocyte growth. The specific activity of cytochrome oxidase in the fraction of heavy mitochondria was twice than in the fraction of light mitochondria. Adenylate cyclase has been localized cytochemically in the zona pellucida, plasma membrane, the smooth membranes of ER, Golgi elements and at the level of yolk platelets. Further correlative histochemical and biochemical studies are needed to determine the changing pattern of enzymes not only in the ooplasm but also in the germinal vesicle during various stages of oocyte growth in the fish. Such studies will be very useful to have an insight into the metabolic changes of growing fish oocyte.

IV. Cortical Alveoli and Yolk

After the various RNA species, organelles and inclusions discussed in Chapter III have accumulated during previtellogenesis, the cortical alveoli and two types of yolk globules named as (1) fatty yolk (or lipid droplets) and (2) proteid yolk (or yolk globules) begin to bc formcd [*Raven,* 1961; *Kraft and Peters*, 1963; *Malone and Hisaoka*, 1963; *Guraya,* 1965, 1982]. Besides the yolk bodies, glycogen may also be present [*Ramadan* et al., 1979b]. With the continued deposition of yolk droplets, the oocyte attains a large size. Actually rapid growth of the oocyte takes place during this later phase of oogenesis, mainly due to the accumulation of cortical alveoli and yolk. This phase is generally called vitellogenesis which is accompanied by intense metabolic activity [*Nakano*, 1969] and is regulated by hormones as will be discussed later on. The rate of oxygen uptake is increased severalfold during vitellogenesis. The incorporation rate of phosphorus ^{32}P and ^{35}S and/or [^{14}C]-labelled amino acids is increased during vitellogenesis. About 80–90% of the amino acids are found in the protein fraction. These observations have indicated that some of the yolk may be synthesized during oogenesis as will be discussed in Section B.

Very divergent views were expressed in regard to the origin, morphology and chemical composition of cortical alveoli and yolk bodies during fish oogenesis. In the classical light microscopic studies on fish oogenesis, their origin was attributed to different cell components such as the nucleoli, yolk nuclei, mitochondria, Golgi apparatus, vacuoles, ground cytplasm, or to a combination of Golgi apparatus, ground cytoplasm and nuclear substances [*Raven,* 1961]. The different views about the origin of cortical alveoli and yolk bodies can be attributed to a variety of factors, but the most important of them appear to be (1) individual interpretations, (2) variations of cortical alveoli and yolk formation in different fish species, and (3) confusion between Golgi bodies, vacuoles and yolk nucleus substance and other inclusions. However, recent correlative electron microscopic, histochemical, biochemical and autoradiographic studies have been very useful to determine the precise origin and nature of cortical alveoli and yolk bodies in fish oocyte. These results will be discussed in detail here.

A. Cortical Alveoli

1. Morphology and Origin

The oocytes of the dogfish *(Scoliodon sorrakowah)* ovary do not show the presence of cortical granules or alveoli comparable to those of amphioxus, cyclostomes, teleosts or amphibians (fig. 29) [*Guraya,* 1982]. The cortical granules have also not been described in the oocytes of other cartilaginous fishes [*Ginsburg,* 1968]. Polyspermy during fertilization of their eggs has been attributed to the absence of cortical granules [*Ginsburg,* 1968].

The large cortical alveoli form the most characteristic feature of the teleost egg (fig. 1b–f, 39) [*Kemp and Allen,* 1956; *T. Yamamoto,* 1961, 1962; *Götting,* 1966; *Hagstrom and Lönning,* 1968; *Ginsburg,* 1968; *Riehl and Schulte,* 1977a; *Donato* et al., 1980; *Takahashi,* 1981]. These have been studied extensively with the techniques of classical cytology, cytochemistry, electron microscopy and autoradiography. The cortical alveoli have also been described as cortical vacuoles, vacuolar yolk, intravacuolar yolk, intravesicular yolk, carbohydrate yolk, yolk vesicles, vacuome, etc. [*Malone and Hisaoka,* 1963; *Guraya,* 1965, 1982; *Ginsburg,* 1968; *Shahi* et al., 1979]. According to most of the investigators on the oocyte growth in teleosts, cortical alveoli make their first appearance in the peripheral ooplasm (fig. 1b, c, 39, 40a, b) from where their formation spreads towards the inner ooplasm (fig. 1d–f, 39, 40c) and finally aggregate in the cortical ooplasm to constitute conspicuous zone (fig. 40d). *Hart and Donovan* [1983] have observed a polarity in the distribution of different sized cortical alveoli in the egg of *Brachydanio rerio,* leading to the establishment of a distinct gradient in the structural organization of the egg cortex from the site of sperm entry to the vegetal pole. The first structures to appear within the oocyte cytoplasm during the gonadotrophin-dependent growth phase are the periodic acid-Schiff (PAS)-positive cortical alveoli [*Konopacka,* 1935]. Actually, the hormonal regulation of cortical alveoli formation is very little studied. In the hypophysectomized goldfish, oestradiol, oestrone and oestriol can induce the formation of yolk vesicles (or cortical vacuoles) [*Khoo,* 1979].

Fig. 39. Histochemical preparation of fish oocytes showing various stages of development and distribution of PAS-positive cortical alveoli (CA). *a* Oocytes of *Puntius ticto. b* Oocytes of *Chana punctatus. c, d* Higher power view of oocytes of *P. ticto. a, b* × 37; *c, d* × 300.

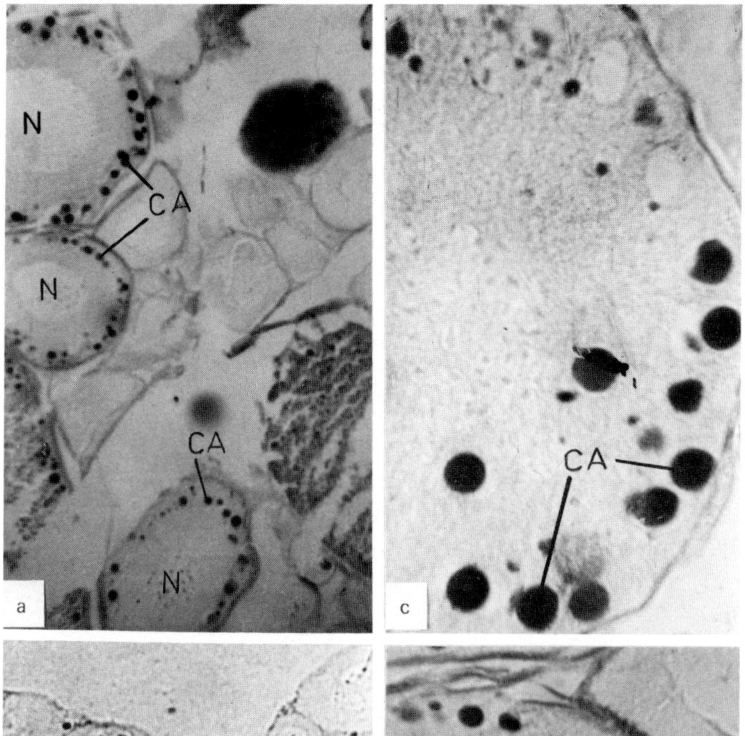
N
CA
N
CA
N
a
CA
c
N
CA
N
b
CA
N
d

Very divergent views about the origin of cortical alveoli in the teleost egg exist in the previous literature. According to *Malone and Hisaoka* [1963] they originate in the pallial substance, which is identical to the basophilic yolk nucleus substance reported by *Guraya* [1963, 1965]; the yolk nucleus substance gives positive histochemical reactions for RNA, protein and very little lipoproteins and corresponds to ultrastructural ribosomes and elements of ER as already discussed in Chapter III. However, *Aketa* [1954] attributed the origin of cortical alveoli to the 'precortex' in the egg of the medaka. Their origin was associated with the 'vacuolar bodies' or ground substance of the ooplasm. In the oocytes of *Puntius ticto,* the cortical alveoli also develop from some vacuolar bodies which gradually accumulate PAS-positive substance in their inside as evidenced from intermediate stages [*Guraya,* 1982]. *Guraya* [1965] using histochemical techniques suggested that the cortical alveoli in *Channa marulius* first appear near the plasma membrane by means of pinocytotic activity. However, a careful examination of growing oocytes of *Channa punctatus* treated with the PAS technique has revealed that small PAS-positive granules first originate in the outer ooplasm (fig. 40a) [*Guraya,* 1982]. The large cortical alveoli are apparently formed by the fusion of these small PAS-positive granules (fig. 40b).

Anderson [1968] using correlative cytochemical and electron microscopy techniques followed the origin of cortical alveoli in the oocytes of teleosts (pipefish and killifish). He suggested that the protein component of cortical alveoli is apparently synthesized by the endoplasmic reticulum and its ribosomes and is subsequently transported to the Golgi complex via vesicles. Within the saccules of the Golgi complex their polysaccharide component is fabricated after which time the Golgi complexes produce vesicles containing the product of the cortical alveoli (fig. 41). The large vesicles containing electron-dense material, which are also described for the teleost *Chrowis chrowis* [*Donato* et al., 1980], appear to correspond to PAS-positive granules described in the oocytes of *C. punctatus* [*Guraya,* 1982]. In the pipefish, in addition to the flocculent component, the vesicles also show one or two eccentrically placed dense structures. When studied in longitudinal section the dense body of a cortical alveolus shows some fine filaments. As the cortical alveolus matures, its internal filamentous components develop a hexagonal pattern. The close morphological relationship of the endoplasmic reticulum with cortical alveoli appears to be indicative of its involvement in the continued synthesis of protein to be utilized by these structures. *Riehl and Schulte* [1977a] observed that the development of

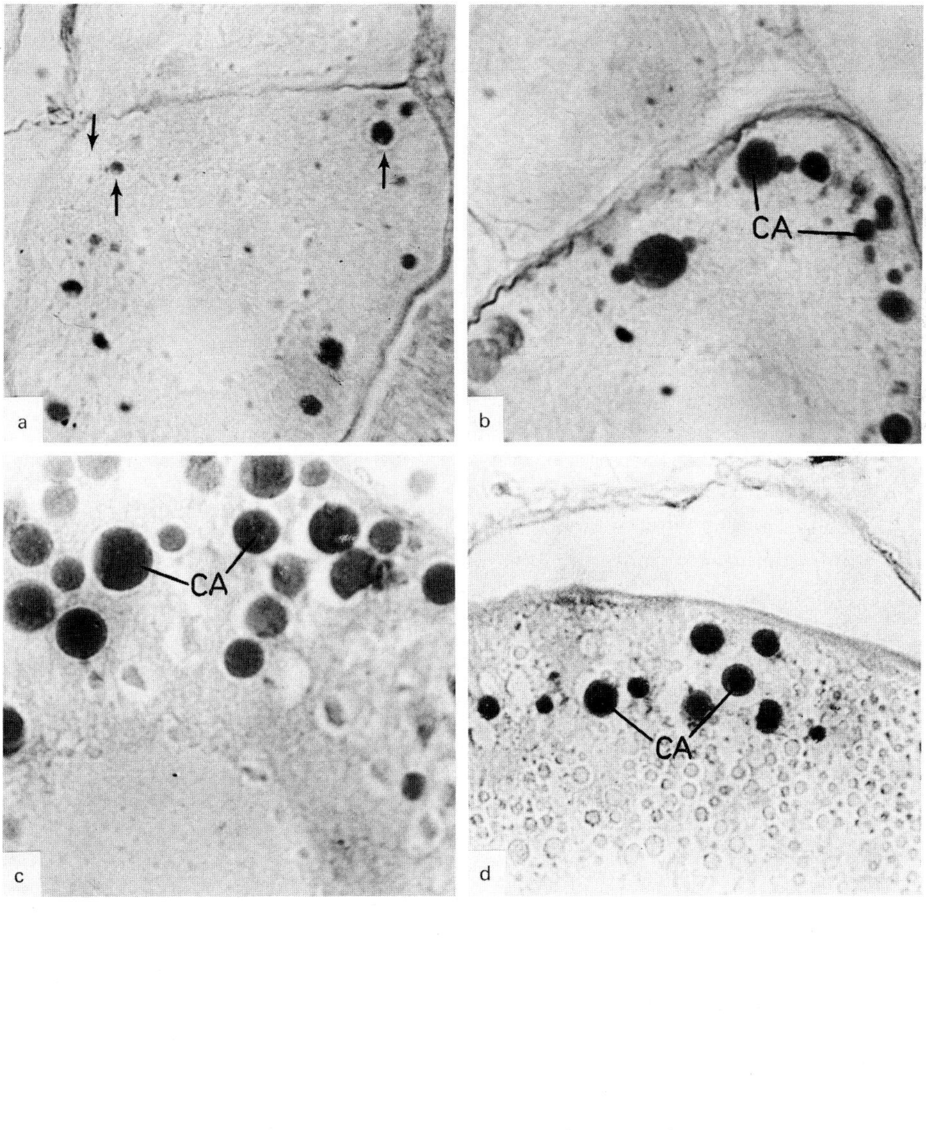

Fig. 40. a The origin of cortical alveoli in the form of PAS-positive granules of various sizes (arrows) in *C. punctatus. b* Growth and development of cortical alveoli by the fusion of PAS-positive granules in the oocytes of *C. punctatus. c* Morphology and distribution of PAS-positive cortical alveoli in the oocyte of *C. punctatus. d* Morphology and distribution of PAS-positive cortical alveoli (CA) in the vitellogenic oocyte of *P. ticto.* Note the more strongly PAS-positive spherical body of cortical alveoli (CA). ×300 [from *Guraya,* 1982].

cortical alveoli in the oocytes of the minnow *(Phoxinus phoxinus)* takes place simultaneously beyond the primary oocyte membrane. These expand centripetally. The cortical alveoli of the minnow derive for the most part from fusion of the vesicles derived from the Golgi complex. According to *Riehl* [1978c] the cortical alveoli in *Gobio gobio* are formed at the end of stage I, whereas in *Noemacheilus barbatulus,* these are developed at the beginning of stage II and originate from the dictyosomes.

Korfsmeier [1966] using autoradiographic techniques observed that the 'intravesicular' yolk spheres (identical to cortical alveoli) within the oocyte cytoplasm of the zebra fish *(Brachydanio rerio)* are labelled soon after the administration of [^{3}H]-histidine and long before the peripheral 'extravesicular' yolk spheres (proteid yolk globules) incorporate labelled material. The cortical alveoli are also labeled by [^{3}H]-glucose, these observations indicate that the cortical alveoli contain a glycoprotein that is synthesized within the oocyte. *te Heesen and Engel* [1973] also identified electrophoretically a glycoprotein which makes its appearance before the accumulation of lipovitelline-like material starts. Immunological studies [*te Heesen,* 1977] further demonstrated that the glycoprotein component is synthesized endogenously within several species of zebra fish. *Tesoriero* [1980] observed that during the earlier time periods, the oocyte of *O. latipes* incorporates more [^{3}H]-glucose than [^{3}H]-galactose in the cytoplasmic and cortical alveoli compartments; overall rate of formation of the cortical alveoli appears to be similar for the 2 tracers. EM autoradiographs have provided more convincing evidence for the participation of cytoplasmic organelles, e.g., ER and Golgi apparatus, in the formation of cortical alveoli. From results of various studies as discussed here, it can be concluded that the cortical alveoli of the fish oocyte are formed by the activity of ER and Golgi complexes; their protein component is synthesized by the ER and then transported to the Golgi complex where it combines with carbohydrates elaborated by the Golgi complex itself. But the regulatory factors as well as molecular mechanisms involved in the formation and maturation of cortical alveoli must be determined in future studies on fish oogenesis.

Electron microscope studies have shown the presence of a 'spherical body' as well as of a colloidal substance within the cortical alveoli of medaka eggs [*M. Yamamoto,* 1964; *Guraya,* 1982]. In certain teleosts, two different populations of cortical alveoli exist as some contain a distinct spherical body [*Droller and Roth,* 1966; *Anderson,* 1968; *Ulrich,* 1969; *Iwamatsu and Ohta,* 1976; *Iwamatou and Keino,* 1976; *Riehl,* 1978c]. Recent transmission and scanning electron microscope studies have shown that the

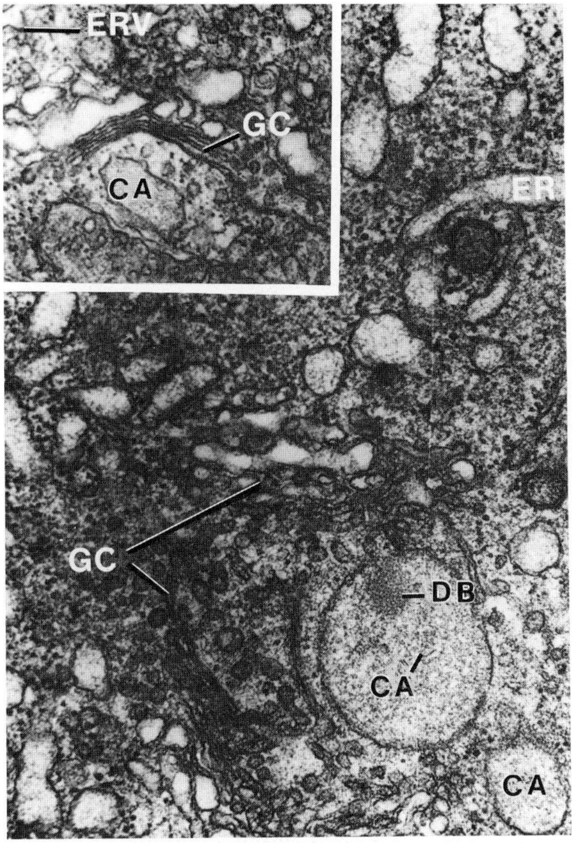

Fig. 41. A section through the Golgi region (GC) of an early stage III oocyte of the pipefish showing vesicles (CA) of varying diameters containing flocculent material (also see inset) and 'immature' cortical alveolus labeled CA. DB = dense body; ER = endoplasmic reticulum. In the inset, note evaginations from the surface of the cisternae of the endoplasmic reticulum (ERV) [from *Anderson,* 1968].

cortical alveolus in the eggs of medaka consists of colloidal material, a spherical body and often a membranous structure [*Iwamatsu and Ohta,* 1976; *Iwamatsu and Keino,* 1978]. Spherical bodies of relatively more dense nature have also been reported in the cortical alveoli treated with the PAS technique (fig. 40d) [*Guraya,* 1982]. Their presence has also been clearly demonstrated in isolated cortical alveoli [*Iwamatsu and Ohta,*

1976]. The spherical body shows a more strong PAS-positive reaction than the surrounding matrix of cortical alveoli (fig. 40d). The source of these two types of substances constituting the cortical alveoli must be determined more precisely in future studies carried out with correlative electron microscope and autoradiographic techniques which have, however, clearly shown that cortical alveoli are formed by the activities of ER and Golgi complex as already stated. The significance of the two populations of cortical alveoli in the fish egg is also not clear.

2. Cytochemistry

Extensive cytochemical studies have been made of the cortical alveoli in the oocytes of different teleosts such as *Pungitius* [*Kusa,* 1953], *Clupea pallasii* [*K. Yamamoto,* 1955a, 1956a, b] *Hypomesus japonicus* [*K. Yamamoto,* 1955b, 1956c], *Liopsetta obscura* [*K. Yamamoto,* 1956d], *Barbus ticto* [*Chopra,* 1958], *Brachydanio rerio* [*Malone and Hisaoka,* 1963], *Channa marulius* [*Guraya,* 1965], *Channa punctatus* [*Shahi* et al., 1979], *Syngnathus fuscus* and *Fundulus heteroclitus* [*Anderson,* 1968]; *Anabas scandens* [*Gopal Dutt and Govindan,* 1969], *Rasbora daniconius* [*Krishnamurthy* et al., 1972], *Noemacheilus barbatulus* and *Gobio gobio* [*Riehl,* 1977c], *Merluccius* [*Ramadan* et al., 1979b] and various other species of teleosts [*Kusa,* 1956; *T. Yamamoto,* 1961; *Ginsburg,* 1968; *Donato* et al., 1980]. Some of these cytochemical studies have also shown that the PAS-positive cortical alveoli develop initially at the periphery of the fish oocyte then show a gradual increase in number and size and meanwhile extend into the central ooplasm (fig. 39, 40a–c) and are finally placed in the cortical ooplasm (fig. 40d). The cortical alveoli continue to show their PAS-positive reaction even after treatment with saliva [*Gopal Dutt and Govindan,* 1969; *Krishnamurthy* et al., 1972]. Such a cytochemical feature of the cortical alveoli has also been reported for *Gobius giuris* [*Rajalakshmi,* 1966] and other fishes investigated by *K. Yamamoto* [1955a 1956b–d]. These studies also revealed that glycogen is absent in them. However, *Krishnamurthy* et al. [1972] demonstrated the presence of glycogen in the cortical alveoli of *R. daniconius* during later stages of oocyte growth. *Shahi* et al. [1979] observed polysaccharides including glyocgen, protein and RNA in the cortical alveoli of *C. punctatus.* These observations in regard to the presence of glycogen and RNA in the cortical alveoli need to be extended and confirmed with correlative cytochemical, autoradiographic and electron microscopic studies. *Anderson* [1968] observed acid mucopolysaccharides and protein in the cortical alveoli (or granules) in *S. fuscus* and *F. heteroclitus,* as also described by *Kusa* [1956]. The PAS-positive substances (carbohydrates) form the most characteristic feature of cortical alveoli during their development and maturation (fig. 39, 40), but no sudanophilic lipids are seen at any stage of their development and maturation (fig. 22b, 23a, b, 42a, b) [*Guraya,* 1965].

Gopal Dutt and Govindan [1969] observed that the cortical alveoli in the oocytes of *A. scandens* contain mucoid-associated acid mucopolysaccharides and do not show any change in their chemical composition during oocyte growth. The cortical alveoli in the oocytes of *C. pallasii* [*K. Yamamoto,* 1956b] and *L. obscura* [*K. Yamamoto,* 1956d] are also not altered in their chemical composition throughout oocyte growth as these continue to show neutral mucopolysaccharides. But in *H. japonicus* the polysaccharide contents of cortical alveoli are altered from natural polysaccharide to mucoid-associated acid mucopolysaccharide [*K. Yamamoto,* 1956c]. The cortical alveoli of *G. giuris* show either galactogen or galactose which are changed into mucopolysaccharides during the maturation of cortical alveoli [*K. Yamamoto,* 1956c; *Rajalakshmi,* 1966]. *Riehl* [1977c] observed that the cortical alveoli of *N. barbatulus* and *G. gobio* contain mostly acid mucopolysaccharides. However, some protein is also present.

From the chemical nature of the polysaccharides demonstrated in cortical alveoli, *K. Yamamoto* [1956a] classified fish eggs into three types: (1) the cortical alveoli showing only neutral polysaccharides (e.g. *C. hallasii* and *L. obscura*); (2) alveoli having at first neutral, but later also acid mucopolysaccharides (e.g. *H. japonicus* and *Salmo*); (3) alveoli containing only acid mucopolysaccharides throughout oogenesis (e.g. *Oryzias*). He has further suggested that these types may be related to the habitat of the fishes, as the first type comprises marine, the second anadromous and the third freshwater fishes. According to this classification of *K. Yamamoto,* the cortical alveoli of freshwater teleost *A. scandens* should have contained acid mucopolysaccharides but these also show mucoids in addition to acid mucopolysaccharides [*Gopal Dutt and Govindan,* 1969].

The cortical alveoli in the oocytes of the freshwater teleost *R. daniconius* are composed of polysaccharide-protein complexes [*Krishnamurthy* et al., 1972] as also reported for *Chrowis chrowis* [*Donato* et al., 1980]; their protein content is rich in cystine and tyrosine whereas the polysaccharide content consists of chondroitin sulphate β and traces of polygalacturonic acid. The chemical composition of cortical alveoli in this fish is also not in conformity with *K. Yamamoto's* classification of eggs. An electrophoretic study of the water-soluble proteins in the dog salmon *(Oncorhynchus keta)* eggs has revealed that the slow-moving band consists of carbohydrate-protein complexes, the mobility of which does not alter after treatment of eggs with isotonic $CaCl_2$ solution [*Guraya,* 1982]. This band appears to represent the contents of cortical alveoli.

Recent studies have demonstrated endogenous lectins within the cortical vesicles (or alveoli) of previtellogenic and early vitellogenic oocytes in several species of fish [*Nosek,* 1984; *Nosek* et al., 1984; *Krajhanzl* et al., 1984a, b]. The multiplication of cortical vesicles and the reduction of their size during oocyte differentiation are accompanied by a strong reaction of their membrane and contents with the antilectin antibody. But it is not known from which structures in the ooplasm the lectin molecules are transported to the cortical vesicles. However, the association of lectins with the vesicle membrane suggests that they may act as specific membrane receptors which direct the polysaccharide molecules to their final destination within the cell [*Krajhanzl* et al., 1984a].

Guraya [1982] has discussed the functional significance of cortical alveoli in the fish egg. Their breakdown at fertilization is closely accompanied by the release of their colloid contents and spherical bodies into the perivitelline space between the chorion (or zona pellucida) and the protoplasmic surface. This, in turn, leads to formation of the perivitelline space which widens rapidly to cause block to polyspermy. *Nosek* et al. [1984] have suggested that fish lectins, which become part of the surface of the fertilized egg, act as a barrier against polyspermy. The precise biological functions of lectins associated with cortical alveoli [*Krajhanzl* et al., 1984a, b; *Nosek,* 1984; *Nosek* et al., 1984] are still not known and should be determined in future studies. The zona pellucida or chorion of fish egg may also be altered during fertilization. The mechanism of initiation of cortical alveolus breakdown is still to be determined more precisely. But a few cortical alveoli located deep within the ooplasm fail to break down [*Iwamatsu and Ohta,* 1976]. The function of cortical alveoli during the cortical reaction at fertilization indicates that these are not yolk in a strict sense but are analogous to the cortical granules that have been extensively studied in the oocytes of numerous invertebrates, protochordates and vertebrates [*Guraya,* 1982]. The cortical granules of animal egg also contain some enzymes, besides the carbohydrates and proteins [*Guraya,* 1982]. The functional significance of these enzymes still needs to be determined at the molecular level during the cortical reaction. The cortical alveoli of fish oogenesis should be studied for their enzymes in order to understand the molecular aspects of cortical reaction in the fish egg.

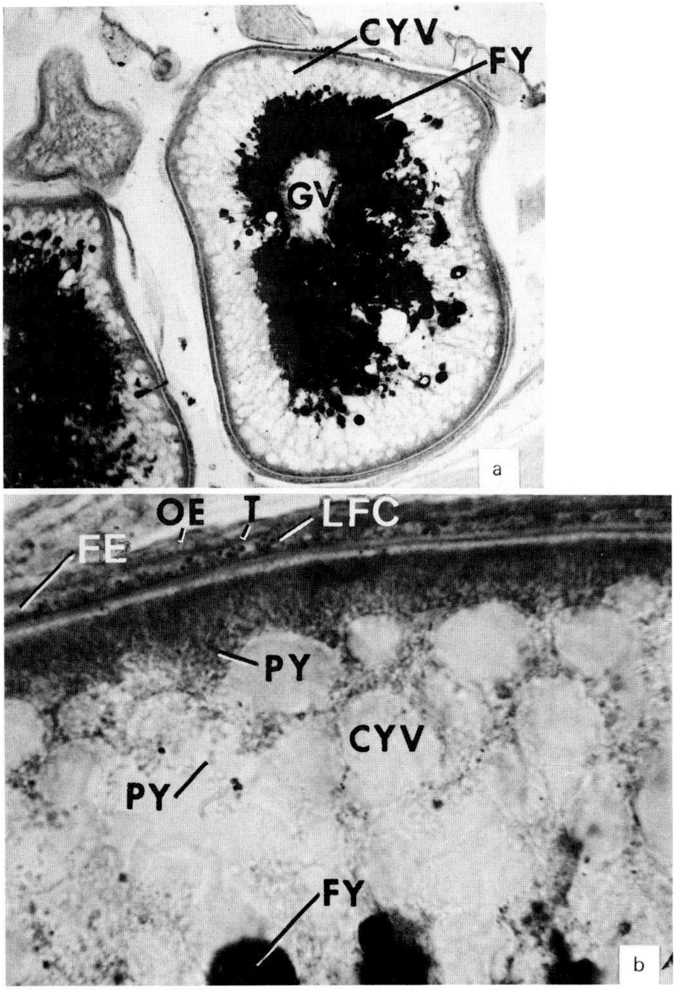

Fig. 42. Photomicrographs of larger oocytes of *Channa,* prepared as in figure 21. *a* The concentration of cortical alveoli (CYV) in the cortical zone, and of fatty yolk (FY) in the central regions of the egg. At this stage, some of the yolk nucleus substance forms a thin layer in the peripheral regions but most of it lies in association with fatty yolk bodies. *b* Different egg membranes, i.e. outer epithelium OE), theca (T), follicular epithelium (FE), zona pellucida (a homogeneous layer) and perivitelline space traversed by striations, and lipid droplets of follicle cells (LFC), small proteid yolk granules (PY) coloring lightly in Sudan black, sudanophobic cortical alveoli (CYV) and peripheral layer of yolk nucleus substance and some mitochondria. *a* × 150; *b* × 810 [from *Guraya,* 1965].

B. Yolk

The dramatic increase in the weight of ovaries of many fishes, from about 1% or less of body weight up to 20% or more, that occurs in the period prior to spawning is mainly due to the accumulation by oocytes of nutrient reserves for the embryos, the yolk. Traditionally, yolk material has been classified by the histochemical reactions of the various ooplasmic inclusions as carbohydrate, lipid and protein yolk [*Raven,* 1961; *Guraya,* 1965]. The yolk in the fish egg is constituted by the fatty (lipid) yolk droplets and proteid yolk globules; the teleost eggs also contain a nongranular fluid yolk. During the terminal stages of oocyte maturation, the egg yolk generally coalesces and mixes with the other egg contents [*Yamamoto and Oota,* 1967a, b; *Goetz,* 1983]. Very divergent opinions were expressed in regard to the origin and nature of yolk bodies in the earlier literature [*Raven,* 1961; *Guraya,* 1965]. From recent investigations carried out with electron microscopy, autoradiography, etc., essentially the following three complex patterns with respect to the origin of yolk precursors in fish oogenesis have merged: (1) that yolk precursors are produced by the oocyte through the conjoined efforts of the organelles and this is called endogenous vitellogenesis; (2) that yolk precursors are fabricated outside the oocyte by another organ, i.e., liver, and are subsequently sequestered within the ooplasm by the process of micropinocytosis and this is called exogenous vitellogenesis; (3) that yolk precursors have a dual origin that encompasses both of the previously mentioned modes. No convincing evidence is yet produced to show that endogenous yolk is actually synthesized by the organelles.

1. Lipid (Fatty) Yolk

Lipid yolk deposition in the fish oogenesis starts prior to that of proteid yolk [*Raven,* 1961; *Guraya,* 1965]. Its appearance can be considered to indicate the beginning of endogenous vitellogenesis [*Shackley and King,* 1977]. Since the major yolk protein precursor, vitellogenin, and its major derivative, lipovitellin, contain lipids, it is pointed out that the classical term 'lipid yolk' is used here for highly sudanophilic inclusions described in morphological studies to be distinct from vitellogenin-derived proteid yolk globules. Fatty yolk consisting mainly of neutral fats (triglycerides) is invariably found in the oocytes of fish [*Raven,* 1961; *Chopra,* 1958a, b, 1960; *Guraya,* 1965; *Ramadan* et al., 1979b]. The exact mode of fatty yolk formation is still unknown. However, *Droller and Roth* [1966] observed that with the start of lipid yolk deposition in the guppy, Golgi cisternae and vesicles circumscribe the lipid droplets that appear in the cortical cytoplasm (fig. 43). Since the cavities of these Golgi elements are still devoid of resolvable material, the precise physiological relationship between the Golgi elements and these lipid bodies is not known and thus should be determined in future studies. The capacity of rainbow trout *(Salmo gairdneri)* ovaries to synthesize triglycerides at a time when vitellogenin uptake is occurring suggests an endogenous source for some of the lipid yolk [*Wiegand and Idler,* 1982]. An exogenous contribution to triglyceride pool from plasma lipoproteins including vitellogenin cannot be ignored [*Leger* et al., 1981]. The precise origin of fatty yolk and the regulatory factors involved in its formation form promising areas for future investigations.

Fatty yolk in the oocytes of *Channa* occurs in the form of highly sudanophilic droplets of various size (fig. 23a, b, 42a, b). The perinuclear concentration of yolk nucleus substance and mitochondria in the oocytes of *Channa* constitutes the first site for the condensation of fatty yolk droplets (fig.23a, b) [*Guraya,* 1965]. Initially, the fatty yolk is in the form of small sudanophilic granules consisting of triglycerides and some phospholipids. With further growth of the oocyte, small fatty yolk granules coalesce to form large fatty droplets. The coalescence of small lipid granules with large lipid droplets has also been observed at the ultrastructural level in the oocytes of guppy [*Droller and Roth,* 1966]. Meanwhile, the large droplets consist of triglycerides only, showing the disappearance of phospholipid component during this transformation [*Guraya,* 1965]. The formation of fatty droplets gradually extends from the perinuclear cytoplasm towards the outer regions of the egg (fig. 23a, b). The process of coalescence further results in the formation of continuous sudanophilic masses of considerable size (fig. 42a) which fill the central regions of larger oocytes. The yolk globules also fuse into a continuous yolk mass in the eggs of other teleosts [*Droller and Roth,* 1966; *Nakano,* 1969]. *Goetz* [1983] has discussed the species variations in regard to the degree of coalescence or fusion of fatty yolk globules.

Gopal Dutt and Govindan [1967] believe that the fatty yolk (L_3 bodies) in the oocytes of *Anabas scandens* derives from lipid bodies designated as L_1 bodies, in collaboration with L_2 bodies, yolk nucleus substance and possibly intranuclear lipids. Lipid yolk globules contain phospholipids besides triglycerides. The oocyte itself forms the seat for lipid synthesis. According to *Riehl* [1977c] during stage II of oocyte growth in *Noemacheilus brabatulus* and *Gobio gobio* neutral lipids (triglycerides) are demonstrated in yolk and cytoplasm. The greater part of lipid yolk in the oocytes of these teleosts is formed by the free cytoplasm, a smaller part is synthesized in dictysomes

[*Riehl,* 1978c]. From the results of various studies carried out so far, it can be suggested that the fatty yolk is deposited in the cytoplasm without much morphological involvement of organelles, except the Golgi bodies or dictyosomes. The physiological meaning of association of dictyosomes with the lipid droplets needs to be determined more precisely in relation to formation of lipid yolk material. The hormonal regulation of lipid yolk deposition in fish oocytes also needs to be worked out.

2. Proteid Yolk

The proteid yolk bodies (or extravesicular yolk), which resist the action of fixing fluids, dehydration and paraffin embedding, are abundantly developed in the eggs of elasmobranchs [*Guraya,* 1978] and teleosts (fig. 16d, e) [*Chopra,* 1958a, b, 1960; *Raven,* 1961; *Guraya,* 1965; *Rastogi,* 1969b; *Guraya* et al., 1975; *Ramadan* et al., 1979b; *Wiegand,* 1982; *Ng and Idler,* 1983]. The proteid yolk granules first originate in the cortical regions of the oocyte and then fill the entire ooplasm. In their size and morphology, the proteid yolk bodies may vary in different regions of the egg.

a. Endogenous Yolk Formation

Various modes for the formation of endogenous proteid yolk have been described in fishes: (a) the yolk is formed in the cytoplasm without any relationship with formed elements [*Raven,* 1961]; (b) in some cases a direct transformation of nuclear emissions into proteid yolk granule has been described [*Raven,* 1961]; (c) proteid yolk granules arise either among the mitochondria or directly from the mitochondria [*Chopra,* 1958a, b, 1960; *Raven,* 1961; *Malone and Hisaoka,* 1963]; (d) the yolk globules often arise in close contact with the mitochondria and yolk nucleus substance [*Guraya,* 1965]. *Ramadan* et al. [1979b] have suggested that mitochondria and Golgi bodies participate in yolk formation in the oocytes of *Merluccius merluc-*

Fig. 43. Portion of guppy oocyte at the onset of lipid yolk formation showing microvilli (mv), follicular epithelium (fe), thecal cells (th), the basement membrane (bm) and large lipid droplets (L) that are surrounded either by Golgi cisternae (G) or by a radially striated cytoplasmic halo (rs) in the oocyte (OOC). Smaller lipid particles (*) from larger inclusions. Mitochondria (m), Golgi elements (G, upper left) are also seen. Inset: Cytoplasm around nucleus (N) is devoid of organelles, except for an occasional mitochondrion

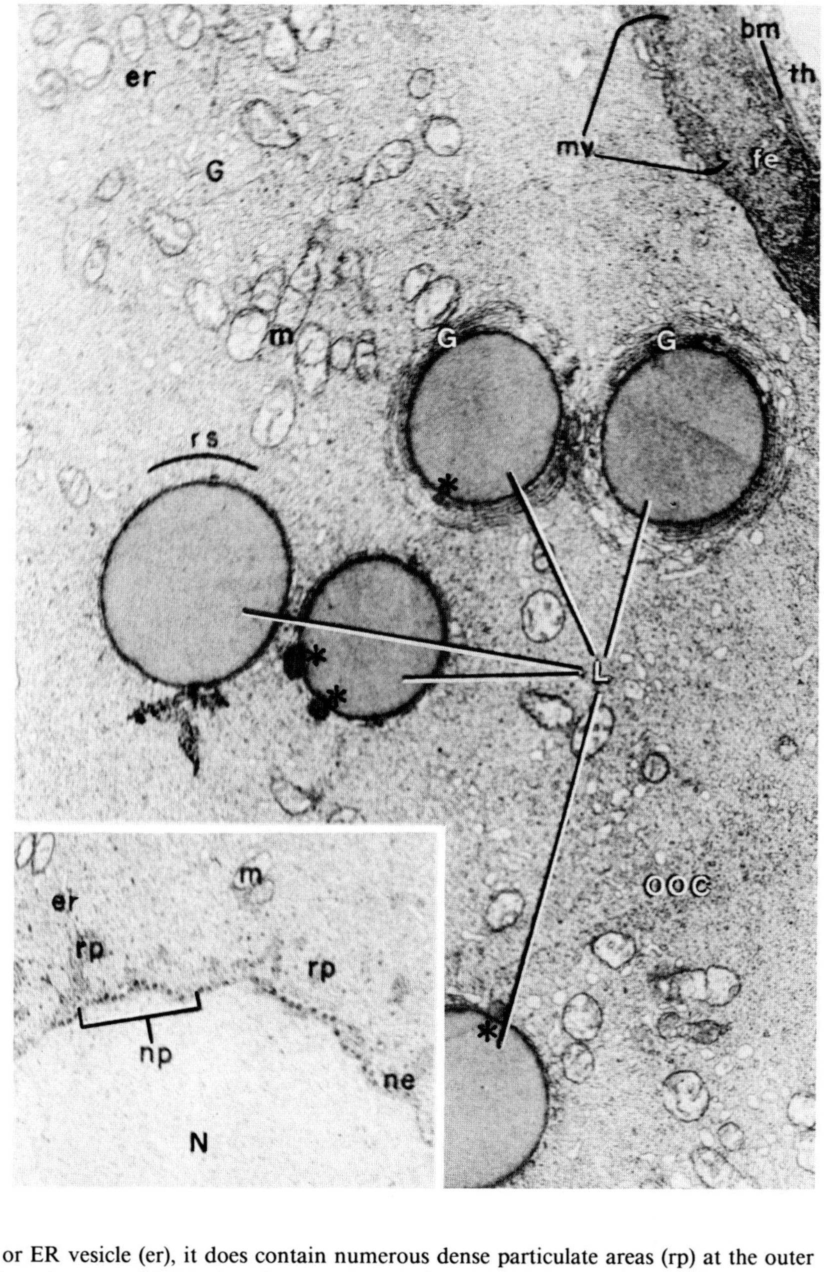

(m) or ER vesicle (er), it does contain numerous dense particulate areas (rp) at the outer surface of pore-studded (np) nuclear envelope (ne). × 16,500; inset × 11,500 [from *Droller and Roth,* 1966].

cius. According to *Kapoor* [1976b] few of the follicle cell nuclei migrate into the ooplasm to form the yolk nucleus which is responsible for yolk deposition in *Puntius ticto* (Ham.). These divergent views about the origin of endogenous yolk can be attributed to species variations as well as to the limited power of light microscope.

Recent electron microscope studies have provided morphological evidence for the involvement of some organelles in proteid yolk formation. *Azevedo and Goimbra* [1978] believe that at the end of dictyate, when the occurrence of Golgi-derived yolk vesicles marks the beginning of vitellogenesis in the viviparous teleost *(Xiphophorus helleri),* mitochondria develop deep infoldings of the limiting membranes, appearing as vacuoles in cross section and containing a fibrillar or flocculent material (fig. 44c, d). The outer limits of the areas occupied by the latter are surrounded by a trilaminar membrane. In other cases, portions of the juxtamitochondrial material appear entirely surrounded by a limiting membrane (fig. 44a, b) making up vesicles of different sizes and apparent fusions of the vesicles originating in mitochondrial infoldings with Golgi-derived yolk vesicles (fig. 44c, d) lead to the formation of large yolk globules in the oocytes of viviparous teleost (fig. 44e, f) *(X. helleri).* These observations have suggested that during vitellogenesis, mitochondria make some contribution to the formation of some components of proteid yolk within recesses of their own contour in the oocytes of *X. helleri.* According to *Riehl* [1978c], the proteid yolk in *N. barbatulus* is formed by transformation of mitochondria and also from multivesicular bodies. But in *G. gobio,* it is formed by the transformation of mitochondria and in annulate lamellae. In *G. gobio,* a part of protein yolk is not surrounded by a visible membrane. It has a crystalline structure and is formed only in stage III. *Riehl* [1977b] demonstrated paracrystalline bodies in young oocytes (stage I) of the loach *(N. barbatulus)* which consist of many osmiophilic particles with a diameter of 15–30 nm. Ultrahistochemical experiments have shown that the paracrystalline bodies are mainly composed of proteins and thus form yolk precursors.

Some electron microscope studies have revealed that there occurs an extensie development of granular ER in early stages of vitellogenesis [*Droller and Roth,* 1966; *Beams and Kessel,* 1973], suggesting its possible involvement in proteid yolk formation. *Droller and Roth* [1966] described in the oogenesis of *Lebistes reticulatus* a twofold synthesis and deposition of protein yolk, one from intraoocyte and the other form extraoocyte sources (fig. 45a, b), as also reported for the teleost *Chrowis chrowis* [*Donato* et al.,

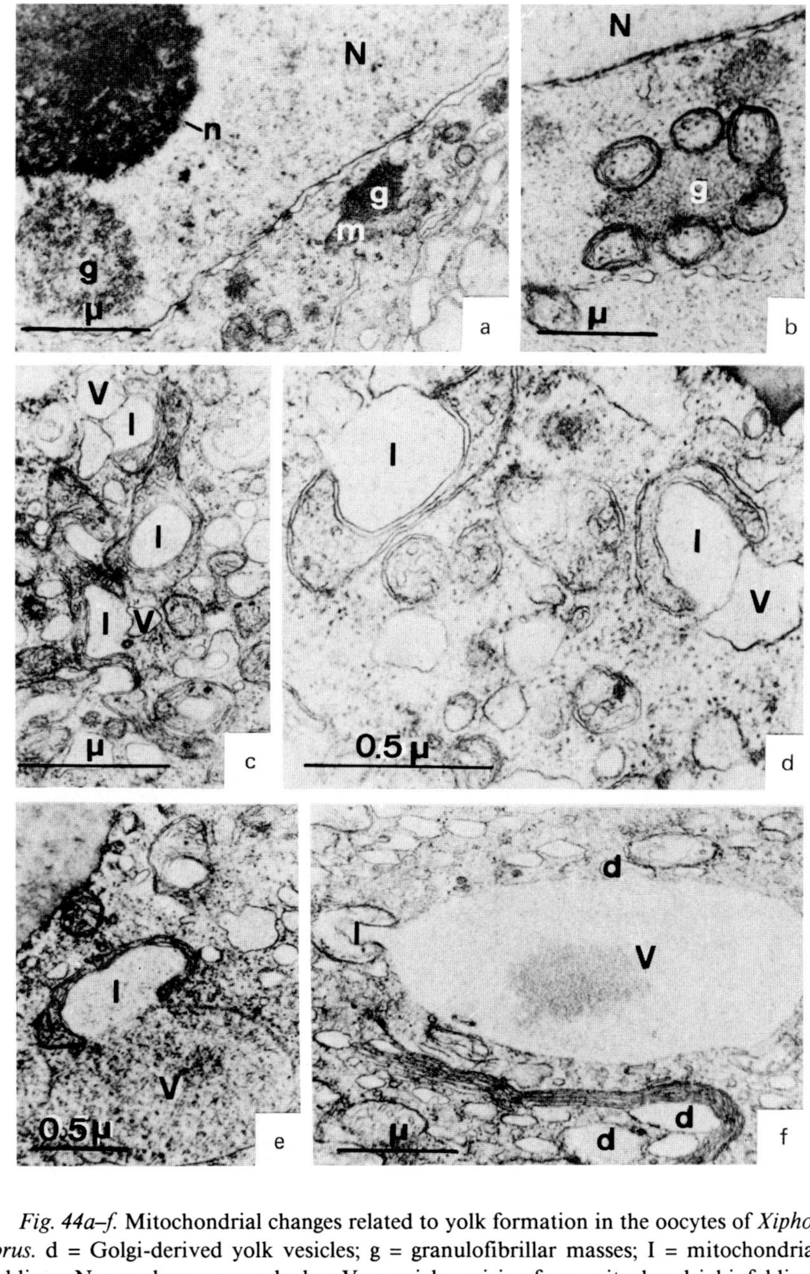

Fig. 44a–f. Mitochondrial changes related to yolk formation in the oocytes of *Xiphophorus.* d = Golgi-derived yolk vesicles; g = granulofibrillar masses; I = mitochondrial infoldings; N = nucleus; n = nucleolus; V = vesicles arising from mitochondrial infoldings [from *Azevedo and Coimbra,* 1978].

1980]. Early in previtellogenesis, smooth vesicles, presumably derived from the endoplasmic reticulum, coalesce to form a yolk droplet containing a somewhat flocculent, loosely packed material within their limiting membrane. In the meanwhile, numerous Golgi vesicles, 70 nm in diameter, are pinched off from the Golgi cisternae, and susequently coalesce to form 'crenate' bodies (fig. 45a). These dissimilar, separately derived droplets then fuse and constitute the intraoocyte contribution to the protein (fig. 45a). Somewhat later in development, the egg appears to ingest extracellular material via bristle-coated micropinocytotic pits and vesicles, 75 and 100 nm in diameter, which apparently fuse to form tubules. The latter then coalesce into larger yolk droplets (fig. 45b). It was postulated that these two different micropinocytotic structures are especially involved with the selective uptake of dissimilar extracellular proteid materials. *Semenov* [1980] also observed that initial stages of yolk inclusion formation in the oocytes of coregonid fishes take place at the expense of autosynthetic processes and during the period of active vitellogenesis, at the expense of heterosynthetic process. *Shackley and King* [1977] indicated an exogenous source for the protein yolk precursors in the oogenesis of the oviparous marine teleost *(Blennius pholis)*. But endogenous contribution to yolk formation is less clear. *Shackley and King* [1979] using tritiated leucine demonstrated an endogenous contribution to protein yolk in isolated oocytes of *B. pholis,* but this may undergo turnover with time. The endogenous yolk formation has been reported in the medium-sized oocytes of zebra fish *(Brachydanio rerio)* [*Korfsmeier,* 1966; *Ulrich,* 1969; *te Heesen and Engel,* 1973].

From the results of various studies as discussed here, it can be concluded that there occurs some endogenous contribution to proteid yolk formation during early vitellogenesis as evidenced from the morphological association of developing yolk granules with the organelles such as mitochondria and vesicles derived from the ER and Golgi complex. But it is still to be determined more precisely whether these organelles actually synthesize the yolk material, or simply act as the site for its condensation. It is also

Fig. 45a. Schematic diagram showing participation of Golgi complexes and rough-surfaced ER in intraoocyte yolk formation in the guppy oocyte. At a later stage (see center of diagram), two types of yolk inclusions coalesce to form large droplets of material comprised of the separate contributions of the ER and Golgi systems to the proteid yolk. *b* The uptake by the developing guppy oocyte of yolk proteins produced elsewhere in the animal

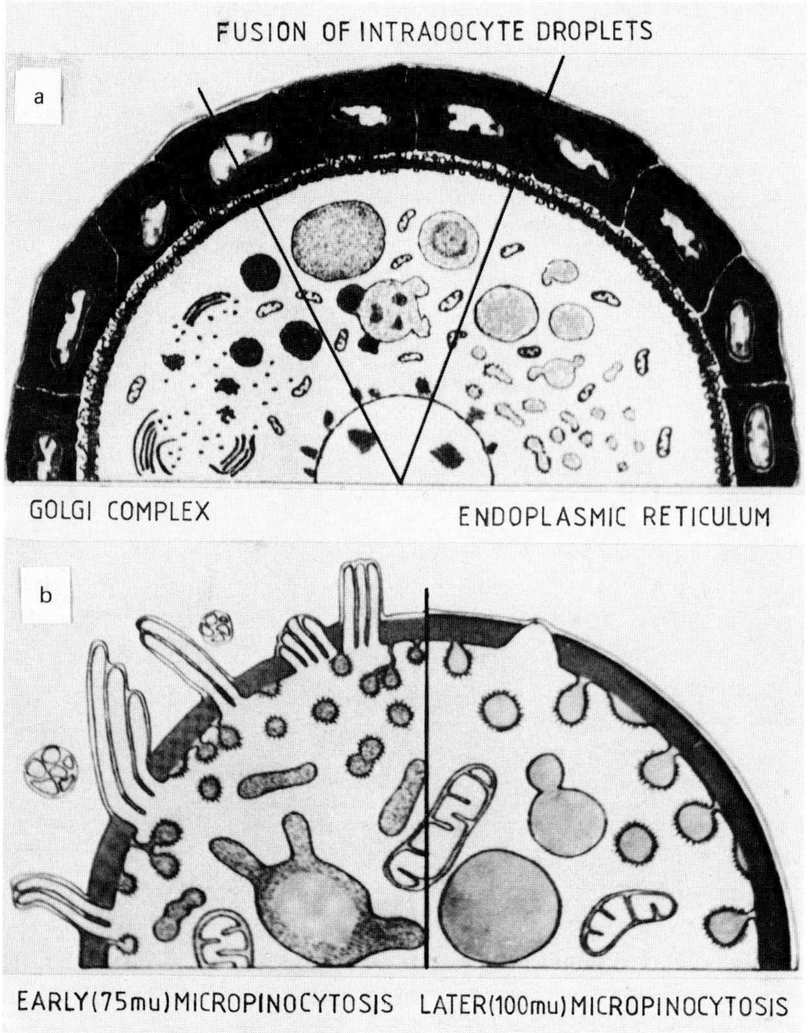

is schematically depicted in this diagram. Numerous 75-μm diameter coated pits and vesicles form at the surface of the egg, become filled with an extracellular granular substance, and pinch off into the oocyte cytoplasm (left side of diagram). These then apparently fuse to form tubules which coalesce into discrete proteid yolk droplets. At a later stage in oogenesis, 100-μm coated vesicles fuse to form larger yolk droplets (right side of diagram) [from *Droller and Roth,* 1966].

possible that some of the ultrastructural descriptions concerning the synthesis and accumulation of putative 'yolk' may be actually describing early cortical alveoli (or yolk vesicles) formation as the process of incorporation of exogenous yolk or vitellogenin into fish oocyte is now well established. Actually the autosynthetic substances do not make a major contribution in this regard. It will be interesting to mention here that in a series of studies on fish gonadotrophins [*Idler and Ng,* 1979], hormonal effects on ovarian metabolism have been studied by measuring uptake of H_3PO_4 and ^3H-leucine in vivo. It is observed that more isotope is incorporated into ovarian proteins which are insoluble in trichloroacetic acid, than can be accounted for by uptake of labelled vitellogenin. It needs to be resolved whether the additional labelled protein is true yolk and whether it is of endogenous or exogenous origin. The synthesis of stage-specific proteins and their regulation at the molecular level form interesting areas for future studies on fish oogenesis.

b. Exogenous Yolk Formation (or True Vitellogenesis)

During recent years greater attention has been paid to the study of various aspects of synthesis, secretion, amounts (levels in the plasma), transport and incorporation of heterosynthetic yolk substances in a number of teleosts [*Wallace,* 1978; *Emmersen and Emmersen,* 1976; *Crim and Idler,* 1978; *Upadhyay* et al., 1978; *Cambell,* 1978; *Borchard,* 1978; *Le Menn,* 1979a, b; *Idler* et al., 1979; *Korsgaard and Petersen,* 1979; *Hara* et al., 1980; *Cambell and Idler,* 1980; *Nath and Sundararaj,* 1981; *Wallace and Selman,* 1981; *Wiegand,* 1982; *Ng and Idler,* 1983], and in several nonteleost fishes [*Pickering,* 1976; *Craik,* 1978a; *Yu* et al., 1980, 1981]. By using immunological, electrophoretic, chromatographic and ultracentrifugal methods as described by *Ng and Idler* [1983] the presence of female-specific vitelloproteins has been demonstrated in their serum, either during their normal breeding period, or in response to administration of oestrogens. Vitellogenin has also been obtained from the blood serum of teleosts under various physiological conditions related to the growth and development of yolky eggs [*Ng and Idler,* 1983; *Ueda* et al., 1984]; the growth and development of vitellogenic eggs is also closely accompanied by the presence of high levels of 17β-oestradiol in the plasma of teleosts [*Kagawa* et al., 1983; *Ueda* et al., 1984; *Yamauchi* et al., 1984; *Hirose* et al., 1985; also see earlier references in *Ng and Idler,* 1983, and *Nagahama,* 1983, 1984]. In vitro studies have indicated that gonadotrophin acts directly on the ovarian follicle to enhance the synthesis and secretion of 17β-oestradiol and its capacity to produce 17β-oestradiol in response to gonadotrophin gradually increases during vitellogenesis, showing a strong close correlation between the plasma levels of this hormone (17β-oestradiol) and vitellogenin. *Cambell and Idler* [1980] made characterization of an oestradiol-induced protein from rainbow trout serum as vitellogenin by the composition and radioimmunological cross-reactivity to ovarian yolk fractions. Biochemical analysis has established similarities between vitellogenin, lipovitellin and phosvitin from trout and other oviparous species [*Wiegand,* 1982; *Ng and Idler,* 1983]. *Craik* [1978a] studied plasma levels of vitellogenin in the elasmobranch *(Scyliorhinus canicula).* In an excellent review, *Ng and Idler* [1983] have discussed the hormonal regulation of vitellogenesis also in different fish species.

Various morphological (including ultrastructural), autoradiographic and biochemical studies have shown that the female-specific protein (vitellogenin) is synthesized and secreted by the liver under the stimulation of oestrogen either of ovarian origin during the breeding season or exogenously administered [*Drilhon and Fine,* 1963; *Krauel and Ridgeway,* 1963; *Thurston,* 1967; *Plack and Fraser,* 1971; *Amirante,* 1972; *Hara,* 1975; *Emmersen and Emmersen,* 1976; *Cambell and Idler,* 1976; *Le Menn and Lamy,* 1977; *te Heesen,* 1977; *Hara and Hirai,* 1978; *Wallace,* 1978; *Larsen,* 1978; *Shackley and King,* 1978; *Hori* et al., 1979; *Korsgaard and Petersen,* 1979; *Takashima* et al., 1979; *Cambell and Idler,* 1980; *Idler and Cambell,* 1980; *Yu* et al., 1980; *de Vlaming* et al., 1980; *Nath and Sundararaj,* 1981; *Wiegand,* 1982; *Ng and Idler,* 1983; *Ueda* et al., 1984], finally found in the blood serum [*Ho and Vanstone,* 1961; *Thurston,* 1967; *Utter and Ridgeway,* 1967; *Aida* et al., 1973; *te Heesen and Engel,* 1973; *Emmersen and Petersen,* 1976; *Shackley and King,* 1978; *Korsgaard and Petersen,* 1979; *Cambell and Idler,* 1980; *Nath and Sundararaj,* 1981; *Ng and Idler,* 1983] and then transported by the blood to the ovaries for its incorporation into the forming yolk spheres of developing vitellogenic oocytes as revealed by autoradiographic studies [*Korfsmeier,* 1966; *te Heesen and Engel,* 1973; *Ng and Idler,* 1983]. Immunological relationships have been shown between vitellogenin and various egg yolk proteins in a number of species and close similarities in amino acid composition between vitellogenin and major yolk protein lipovitellin have been observed [*Hara and Hirai,* 1978; *Cambell and Idler,* 1980; *Ng and Idler,* 1983]. *Hara* et al. [1980] have suggested vitellogenin in eel may be incorporated into yolk without modification. *Hori* et al. [1979] also made a similar suggestion in regard to vitellogenin in *C. auratus,* but *de*

Vlaming et al. [1980] have not agreed to this suggestion as they have isolated lipovitellins and phosphovitins from *C. auratus,* showing that typical proteolytic metabolism of vitellogenin takes place upon uptake by the oocyte in this species.

Peute et al. [1978] suggested that ultrastructural changes in the hepatocytes of the zebra fish *(Brachydanio rerio),* which coincide with special phases of ovarian activity, are indicative of vitellogenin synthesis in the liver [see also *Upadhyay,* 1977]. Steroids, especially 17β-oestradiol, may trigger this process. The cell and molecular biology of liver cells, which is indicative of protein synthesis (or vitellogenin) in the liver, has been discussed in some recent reviews [*Ng and Idler,* 1983; *Fostier* et al., 1983]. In response to endogenous or exogenous oestrogen, the liver is increased in size and its cells show the enlargement of the Golgi complex and proliferation and accumulation of abundant granular ER and numerous ribosomes related to protein synthesis as evidenced from the incorporation of labelled amino acids. Hepatic RNA content is increased. Vitellogenin mRNA has also been isolated from the liver of some teleosts. From the results of various studies, it is now well established that concomitant with the increased activity of the liver is the enhancement in the steroidogenic activity in the ovarian follicular envelope which is regulated by pituitary gonadotrophin (or vitellogenic gonadotrophin).

Comparisons of the vitellogenic potency of 3 oestrogens in ovariectomized *Tilapia aurea* have shown that oestradiol is more potent than oestriole which, in turn, is more potent than oestrone [*Terkatin-Shimony and Yaron,* 1978]. *Korsgaard and Petersen* [1979] observed that oestradiol treatment during pregnancy in the blenny *(Zoarces viviparus)* results in a dose-dependent increase in vitellogenin and lipids of the blood. In excellent reviews, *Wiegand* [1982] and *Ng and Idler* [1983] have discussed the roles of various sex steroids in relation to vitellogenesis in teleosts, elasmobranchs and cyclostomes.

From this discussion it can be concluded that oestrogen induction of vitellogenin synthesis in the liver has been established now. Meanwhile, there also occur other metabolic changes in the liver and plasma of the female [*Ng and Idler,* 1983]. The various studies have clearly demonstrated a significant correlation between the stage of ovarian development and the plasma vitellogenin levels, the plasma oestrogen concentration and the gonadosomatic index, revealing the involvement of oestrogen and vitellogenin in ovarian development and maturation and in the increase of ovarian size [*Idler* et al., 1981; *Ueda* et al., 1984].

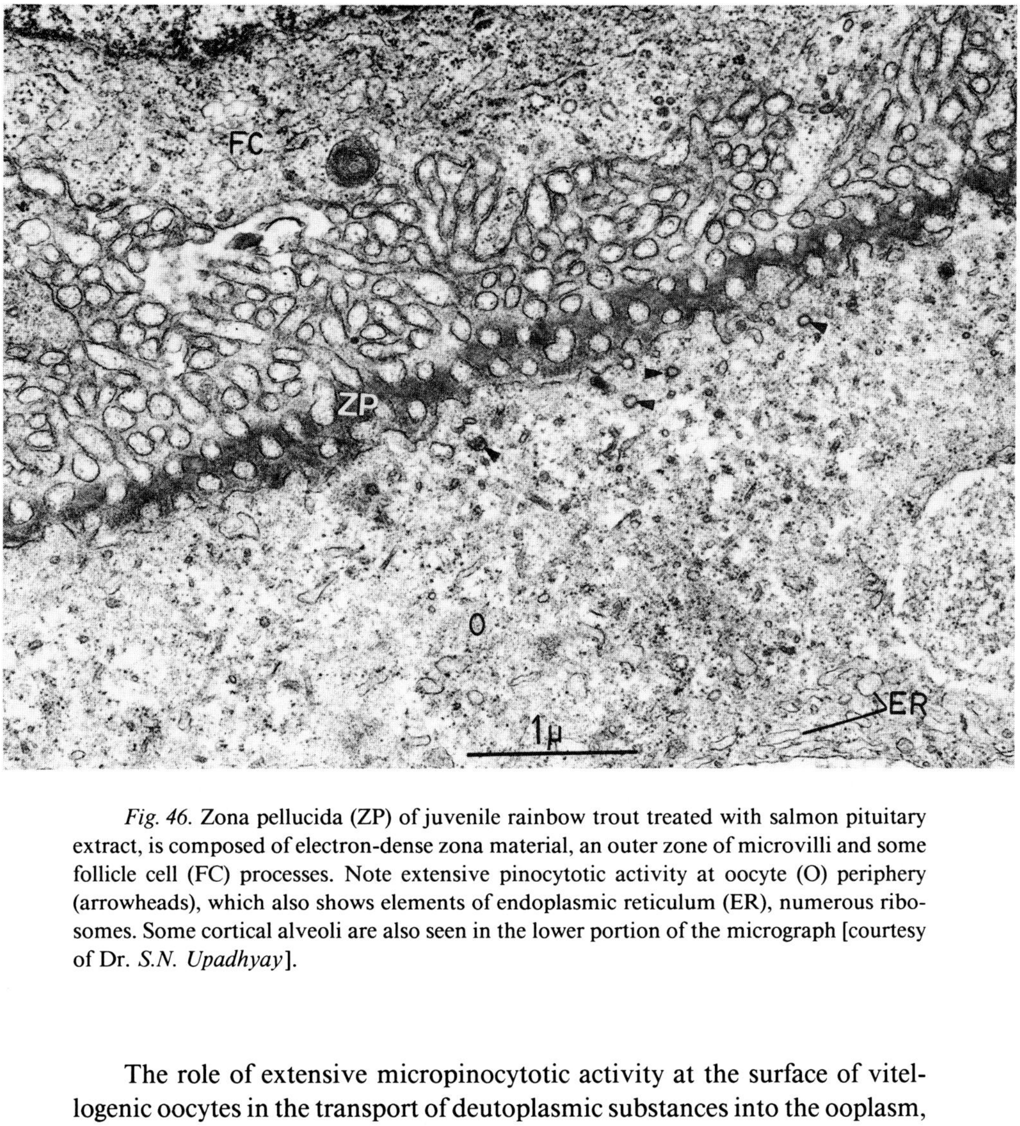

Fig. 46. Zona pellucida (ZP) of juvenile rainbow trout treated with salmon pituitary extract, is composed of electron-dense zona material, an outer zone of microvilli and some follicle cell (FC) processes. Note extensive pinocytotic activity at oocyte (O) periphery (arrowheads), which also shows elements of endoplasmic reticulum (ER), numerous ribosomes. Some cortical alveoli are also seen in the lower portion of the micrograph [courtesy of Dr. *S.N. Upadhyay*].

The role of extensive micropinocytotic activity at the surface of vitellogenic oocytes in the transport of deutoplasmic substances into the ooplasm, at the time of yolk deposition, has been demonstrated by electron microscopy for several species of teleosts (fig. 45, 46) [*Droller and Roth,* 1966; *Anderson,* 1968; *Wegmann and Götting,* 1971; *Upadhyay,* 1977; *Upadhyaya* et al., 1978; *Selman and Wallace,* 1982a; *Ng and Idler,* 1983]. *Droller and Roth* [1966] observed that the size of bristle-coated micropinocytotic pits and vesicles is somewhat larger in late stages of vitellogenesis than in early stages. From these electron microscope studies, they have suggested that two different micropinocytotic activities are specifically involved in the selective uptake of dissimilar extracellular materials (fig. 45a, b). *Selman and Wallace* [1982a, b] have used the electron-dense tracer, horseradish peroxidase, and a larger heterologous protein, *Xenopus laevis* [^{3}H]-vitellogenin, to demonstrate the pathway by which exogenous proteins leave the maternal circulation, and pass through the follicle to yolk spheres within the growing oocytes of teleosts. Proteins leave the perifollicular capillaries, and traverse the theca, the follicular epithelium and the vitelline envelope (zona pellucida). At the oocyte surface these are incorporated via micropinocytosis, substantiating the significance of heterosynthetic processes during oocyte growth. The clearance from the plasma of phosphoprotein (an index of vitellogenin) and its appearance in yolk granules has been shown in the elasmobranch *(S. canicula)* [*Craik,* 1978b].

The electron microscope studies have shown that substance incorporated by micropinocytosis is transferred to forming yolk spheres within the peripheral ooplasm [*Droller and Roth,* 1966; *Anderson,* 1968; *Ulrich,* 1969; *Wegmann and Götting,* 1971; *Gupta and Yamamoto,* 1972; *Shackley and King,* 1977; *Selman and Wallace,* 1982a, b; *Ng and Idler,* 1983]. These forming fully formed proteid yolk spheres can also be studied in routine histological preparations (fig. 47a–d). In the central regions of the egg, the fusion of proteid yolk globules can also be seen (fig. 47c, d). Crystalline yolk has been reported in the granules of several distantly related teleosts [*Yamamoto and Oota,* 1967; *Ulrich,* 1969; *Gupta and Yamamoto,* 1972; *Flegler,* 1977; *Shackley and King,* 1977; *Riehl,* 1978c; *Lange* et al., 1982]; in some teleosts the yolk globule consists of a crystalline main body surrounded by a superficial layer and outer membrane [*Yamamoto and Oota,* 1967]. The zonation of proteid yolk spheres can also be observed with the light microscope as their peripheral regions stain more intensely than the central region (fig. 47b). In most teleosts, however, yolk proteins accumulate in fluid-filled yolk spheres [*Grodzinski,* 1954, 1973] rather than crystalline granules. These yolk bodies may either continue their integrity through-

---

Fig. 47. Vitellogenic oocytes of *Cyprinus carpio,* showing variable morphology of proteid yolk platelets (PY) in different regions of ooplasm. *a* The distribution of proteid yolk platelets (PY), nucleus (N) and follicular wall (FW). *b* Proteid yolk platelets of variable

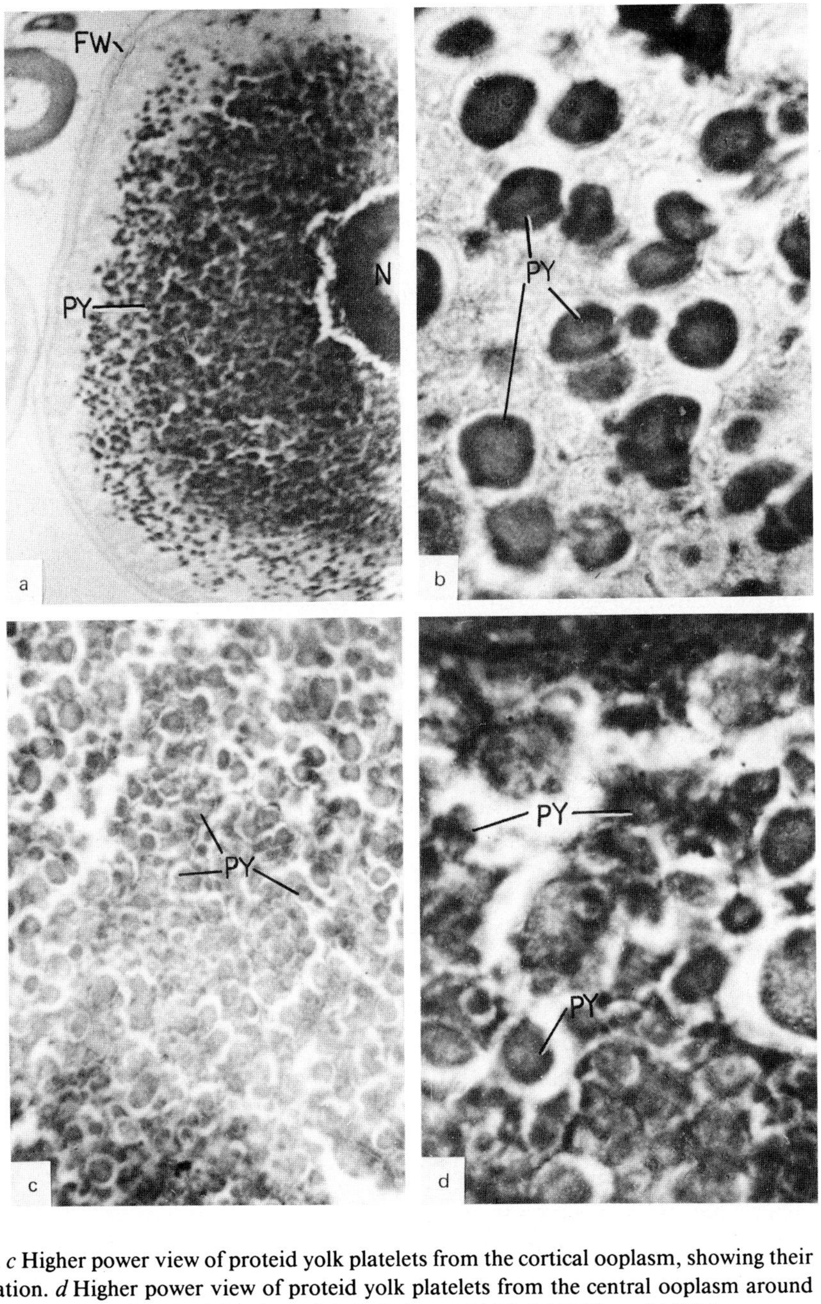

size. *c* Higher power view of proteid yolk platelets from the cortical ooplasm, showing their zonation. *d* Higher power view of proteid yolk platelets from the central ooplasm around the nucleus, showing the initiation of their fusion. *a* × 75; *b* × 300; *c, d* × 750.

out the oocyte growth, or fuse centripetally eventually to constitute a continuous mass of fluid yolk, a process which confers on many teleost eggs their specific transparency [*Wallace and Selman,* 1980; *Goetz,* 1983; *Nagahama,* 1983]. This process may occur relatively soon after the initial development of spheres during the later stages of vitellogenesis, or during maturation. As a result of these changes, the ooplasm is displaced into a peripheral rim surrounding the yolk mass. The biophysical factors involved in the fusion of yolk bodies need to be determined.

The vitellogenic protein produced and secreted by the liver of the female under oestrogen stimulation is transported via the blood to the ovaries under pituitary mediation [*Cambell and Idler,* 1976; *Cambell,* 1978; *Crim and Idler,* 1978; *Wallace,* 1978; *Ng and Idler,* 1978, 1983]. Gonadotrophin-sensitive and selective uptake of *S. gairdneri* oocytes is shown in vitro [*Cambell,* 1978; *Cambell and Jalabert,* 1979]. Autoradiography of oocytes from sheepshead minnow, *Cyprinodon variegates,* which were given injections 24 h earlier with tritiated *Xenopus* vitellogenin, showed grain deposition over yolk spheres and fusing yolk masses, indicating incorporation of the heterologous vitellogenin into yolk [*Selman and Wallace,* 1982a, b]. *Ng and Idler* [1980] injected female winter flounder *(Pseudopleuronectes americanus)* with labelled serum phosphoprotein obtained from oestrogen-treated males and showed gonadotrophin-sensitive uptake of label into yolk. Immunofluorescence has established that vitellogenic gonadotrophin binds to ooplasm of yolky oocytes and to the follicular envelope supporting the suggestion that the hormone acts on the yolky oocytes to stimulate incorporation of vitellogenin [*Ng and Idler,* 1983]. But the mechanisms involved need to be determined at the molecular level. The more recent correlative ultrastructural and physiological studies have further indicated that salmon pituitary extract (10 μg/g) induces the incorporation of exogenous yolk into the oocyte as evidenced by an extensive pinocytotic activity at the oocyte surface (fig. 46) [*Upadhyay,* 1977; *Upadhyay* et al., 1978]. These observations have suggested that gonadotrophin is necessary for the induction of endogenous yolk synthesis and that some other pituitary hormone(s) is involved in the process of incorporation of exogenous yolk or vitellogenin into the oocyte. The pinocytotic activity is believed to be initiated by the pituitary hormone(s) [*Upadhyay,* 1977; *Upadhyay* et al., 1978]. Meanwhile, parallel increases occur in the accumulation of yolk lipophosphoprotein (phosvitin) in the ovary [*Crim and Idler,* 1978]. The nature of the vitellogenic material in the fish pituitary needs to be determined in future investigations [*Crim and Idler,* 1978; *Ng and Idler,* 1983]. *Cambell*

[1978] observed that follicles, isolated from ovaries of rainbow trout before the onset of active vitellogenesis, incorporate [^{14}C]-labelled vitellogenin when incubated in vitro for 18 h. An aqueous extract of lyophilized salmon pituitaries stimulates the incorporation of vitellogenin, but not other serum proteins. A pituitary fraction, which is not bound to Con A Sepharose (during two chromatographies) stimulates incorporation in a similar fashion to the total extract. Less than 20% of the total radioactivity incorporated into the follicles is associated with follicle cells and removal of these tissues after incubation has revealed that hormonal treatment results in a real stimulation of vitellogenin incorporation into the oocytes. These results further support the suggestion that the incorporation of vitellogenin by trout oocytes may be under direct control by a pituitary gonadotrophin [see also *Ng and Idler,* 1983].

Where investigated, vitellogenin uptake into oocytes has been observed to be highly selective [*Wallace and Jared,* 1976; *Wallace* et al., 1980], although the available evidence for teleosts shows a reduced specificity [*Cambell and Jalabert,* 1979]. In addition, electron microscope studies on the swordtail, *Xiphophorus helleri,* have demonstrated incorporation of an electron-dense marker into the yolk bodies of vitellogenic oocytes [*Wegmann and Götting,* 1971]. This indicates that the uptake of nonvitellogenin macromolecules can also occur to some extent. There are also other reports suggesting the possibility of uptake of proteins other than vitellogenin from the blood [see review by *Ng and Idler,* 1983]. The oocytes of *S. gairdneri* can sequester nonvitellogenin serum proteins in vitro [*Cambell and Jalabert,* 1979]. Uptake of these proteins is not sensitive to gonadotrophin and occurs at a much lower rate than uptake of vitellogenin. *Aida* et al. [1973] also observed the presence in the egg of a protein common to both males and females and thus not strictly a vitellogenin. A similar protein common to male, nonvitellogenic female and vitellogenic female hagfish *(Eptatretus stouti)* plasma has been observed using an antiserum to egg yolk [*Yu* et al., 1980].

c. Histochemistry

Various histochemical studies have shown that protein yolk bodies in the fish egg consist mainly of protein and lipoproteins [*Chopra,* 1958a, b, 1960; *Malone and Hisaoka,* 1963; *Lal,* 1963; *Guraya,* 1965; *Rastogi,* 1969; *Shahi* et al., 1979]. The proteid yolk platelets in *Channa* also contain carbohydrates [*Guraya,* 1965]. The proteid yolk of *Noemarcheilus barbatulus* and *Gobio gobio* consists of proteins, lipids and polysaccharides [*Riehl,* 1977c]. The proteid yolk platelets in *N. barbatulus* are zonated as evidenced by their three layers. The central portion consists predominantly of proteins and lipids while the two lateral portions are mostly composed of proteins and acid mucopolysaccharides. The lipoprotein crystals have been shown in the yolk platelets of a teleost, *Pelvicachromis pulcher* [*Lange,* 1981]. Further correlative histochemical and electron microscope studies are needed to reveal the molecular organization of yolk platelets in fish egg.

Acid phosphatase activity is localized at the periphery of yolk globules in the egg of the teleost *(Catostomus commersoni)* [*Lessman,* 1981]. The use of combined histoenzymological, electrophoretical and spectrophotometrical methods in the identification of lysosomal enzymes in the oogenesis of *Carassius auratus* has enabled the consideration of the vitelline platelets as a lysosomal structure [*Varo* et al., 1977]. Oocytic lysosomes occur as small granules, first under the plasmalemma, then, as the cortical alveoli arrange themselves, at the oocyte periphery. Finally they spread in the entire cytoplasm. Meanwhile, the vitelline platelet is formed by the emergence of a phosphoprotein crystalloid in a vesicle with acid phosphomonoesterase, arylsulfatase, arylesterase and β-glucuronidase.

Riehl [1980a] demonstrated ATPase in all four oocyte stages of *Heterandria formosa* but its level is highest in stages II and III in which yolk formation takes place, suggesting its role in the transport of yolk material into the oocyte. Na^+-, K^+-activated ATPase is observed on the membranes of both the yolk vesicles and the cortical granules and on the oocyte and follicular microvilli and the primary oocyte membrane.

d. Biochemistry

The biochemical studies have demonstrated a variety of chemical components such as the proteins, carbohydrates, glycoproteins, lipids, carotenoids, etc., either in the whole ovaries or in isolated eggs. But these chemical components are believed to derive mainly from the yolk of fish egg as also supported by histochemical data. Therefore, vitellogenin and egg yolk derivatives have been investigated in some detail in a number of teleost species. M_r determinations by gel filtration [*Emmersen and Petersen,* 1976; *Hara and Hirai,* 1978; *Cambell and Idler,* 1980; *Hara* et al., 1980] and electrophoresis [*Hori* et al., 1979; *de Vlaming* et al., 1980; *Cambell and Idler,* 1980; *Ng and Idler,* 1983] have demonstrated values between 326,000 and 600,000 for the various intact molecules. Actually variable M_rs have been reported for vitellogenins of different teleost species [*Ng and Idler,* 1983]. During recent years, there has been a considerable interest in the isolation and characterization of yolk constituents in different fish species

[*Wiegand,* 1982; *Ng and Idler,* 1983]. The various methods used in the isolation purification and characterization of yolk components are described in an excellent review by *Ng and Idler* [1983] and thus will be omitted here.

Proteins. The yolk proteins isolated from the eggs of teleosts are generally atypical. M_rs of delipidated polypeptide monomers studied by electrophoresis in the presence of sodium dodecyl sulphate (SDS) range from 140,000 to 220,000 [*Hara and Hirai,* 1978; *Hori* et al., 1979; *de Vlaming* et al., 1980; *Roach and Davies,* 1980]. The monomeric and dimeric forms of vitellogenin along with a third, comparatively minor component of intermediate M_r have been reported for *C. auratus* with native gel electrophoresis [*Hori* et al., 1979; *de Vlaming* et al., 1980]. The latter authors have further observed multiple M_r forms of delipidated vitellogenin monomer.

The lipovitellins (or glycolipophosphoproteins) of yolk which are usually heterogeneous, show very low contents of alkali-labile phosphorus unlike all other lipovitellins; these are generally soluble in solutions of low ionic strength [*Jared and Wallae,* 1968]. *de Vlaming* et al. [1980] have reported two M_r classes of lipovitelline subunits from *C. auratus,* each of which consists of more than one polypeptide. The presence of high and low M_r subunits has been demonstrated using a 2-mercaptoethanol-reduced preparation of lipovitellin from *S. gairdneri* [*Hara and Hirai,* 1978]. Trout lipovitellin shows a M_r ranging from 244,000 to 300,000 [*Ng and Idler,* 1983]. Low or undetectable levels of protein phosphorus are described for lipovitellins in a number of teleost species [*Jared and Wallace,* 1968; *Markert and Vanstone,* 1971; *Plack* et al., 1971; *Cambell and Idler,* 1980]. A lipid content of 25% and an alkali-labile phosphorus content of 0.007% are described for trout lipovitellin [*Cambell and Idler,* 1980]. The lipids of lipovitellin will be discussed later on in the subsection on lipids. Two types of lipovitellin, which can be partially resolved by DEAE-cellulose chromatography, have been reported for the cod ovary [*Plack and Fraser,* 1971]. The trout lipovitellin consists of two subunits [*Ando,* 1965]. The M_rs of its two subunits in the case of goldfish range between 105,000–110,000 and 19,000–25,000 [*de Vlaming* et al., 1980].

The phosphoprotein derivatives of vitellogenin, which are called phosvitins, have been studied extensively in different teleost species and a M_r of 19,000–43,000 [*Ng and Idler,* 1983]. These proteins may be isolated by redissolving a yolk dilution precipitate in saline and precipitating lipovitellin (and B-component) with ammonium sulphate [*Wallace* et al., 1966; *Markert and Vanstone,* 1971]. Further purification of phosvitin can be obtained with chromatography. Other methods have also been used for its purification [*Mano and Lipmann,* 1966; *Mano and Yoshida,* 1969; *Inoue* et al., 1971; *Ng and Idler,* 1983]. Fish ovaries show variations in regard to phosvitin content [*Ng and Idler,* 1983]. Very little phosvitin is found in the ovaries of several marine teleosts, and their developing eggs may obtain their required phosphorus for metabolic purpose from the environment [*Ng and Idler,* 1983]. All these studies have shown that the phosvitins are frequently heterogeneous showing multiple forms with varying degree of phosphorylation [*Mano and Lipmann,* 1966; *Jared and Wallace,* 1968] and show lower molecular weights [*Schmidt* et al., 1965; *Wallace* et al., 1966; *Markert and Vanstone,* 1971] and have widely divergent amounts of protein phosphorus [*Barman* et al., 1964; *Mano and Lipmann,* 1966]. Sometimes, phosvitins may be completely absent from yolk preparations of some marine teleosts as already stated [*Jared and Wallace,* 1968]. The various fractions of phosvitin show the same amino acid composition and are thus believed to be produced by stepwise dephosphorylation and phosphorylation in the course of vitellogenesis [*Ng and Idler,* 1983]; protein kinases play an important role in the phosphorylation of phosvitin. But salmon phosvitin gives a single peak on DEAE-cellulose. The results of various studies have shown that phosvitins are rich in phosphorus but do not show lipid and carbohydrate; the phosvitin shows an alkali-labile phosphorus content of 15.8% [*Cambell and Idler,* 1980]. Some glucosamine has been observed in salmonid phosvitin [*Mano and Yoshida,* 1969].

Amino acids of phosvitins have been studied. Serine is present predominantly in phosvitins of teleosts [*Ng and Idler,* 1983]. Its proportions range from 42% in *S. gairdneri* [*Cambell and Idler,* 1980] to 75% in *C. pallasii* [*Inoue* et al., 1971]. However, serine contents of phosphoproteins isolated from shad range from 9 to 22% [*Mano and Lipmann,* 1966]. The phosvitins are poor generally in histidine, sulphur-containing and aromatic amino acids [*Mano and Lipmann,* 1966]. This observation is verified in subsequent studies [*Inoue* et al., 1971; *Cambell and Idler,* 1980]. The ester linkage of phosphate to serine has been investigated in *C. pallasii* [*Inoue* et al., 1971]. In the salmonid phosvitin, tyrosine is the NH-terminal residue and aspartic acid forms the C-terminal residue [*Mano and Yoshida,* 1969]. Phosvitin is believed to form a reservoir of phosphorus and is, therefore, an energy source for embryonic development [*Ng and Idler,* 1983]. However, it is not found in the eggs of some marine teleosts. The phosvitins of *O. kisutch* [*Markert and Vanstoni,* 1971] and *S. gairdneri* [*Cambell and Idler,* 1980] do not show demonstrable contents of lipid.

Some of the chemical components obtained from yolk preparations do not appear to show the characteristics of either lipovitellin or phosvitin [*Jared and Wallace,* 1968], and at least one of these, the 'B component' observed in salmonids, does not show lipid or protein phosphorus; its M_r ranges from 21,000 to 30,000 [*Ng and Idler,* 1983] and it derives from the serum of sexually maturing females [*Markert and Vanstone,* 1971]. These many anomalies led *Jared and Wallace* [1968] and *Mano and Lipmann* [1966], to suggest that teleost yolk proteins are phosphorylated during synthesis in either a deficient or heterogeneous manner. In view of more recent information on (1) the nature and origin of the B component in salmonids [*Markert and Vanstone,* 1971]; (2) the extreme susceptibility of at least one teleost vitellogenin to proteolysis [*Hickey and Wallace,* 1974], and (3) the derivation of yolk proteins from vitellogenin by proteolysis [*Bergink and Wallace,* 1974], it has been suggested that teleost vitellogenins are proteolytically cleaved at alternate and/or more numerous sites during their conversion to yolk proteins [*Wallace,* 1978]. An immunological relationship between B component and vitellogenin is shown in *S. gairdneri* suggesting that B component is a part of vitellogenin complex [*Cambell and Idler,* 1980] However, various characteristics of B component show that it is a protein different from vitellogenin, lipovitellin and phosvitin and thus believed to be a heterosynthetic or an autosynthetic protein [*Cambell and Idler,* 1980]. It is suggested here that the role of germinal vesicle in this regard should be investigated as it shows extensive accumulation of protein granules in its nucleoplasm and perinuclear ooplasm during vitellogenesis (fig. 1, 11, 16). The contribution of these protein granules to proteid yolk formation deserves special attention in future investigation as such protein granules are not seen in the germinal vesicle of previtellogenic oocytes.

Bussmann [1977/1978] has obtained acidic, strong basic and weak basic heat-resistant proteins and polypeptides from the ethanol extracts of the homogenized ovaries of Baltic herring (*Clupea harengus*), which have been divided into maturity stages IV, V and VI. Alterations in protein composition are observed during these three stages. The content of proteins soluble in the phosphate buffer in the mature eggs of the bream is higher than those of the roach [*Konovalov and Zhukinskii,* 1979]. In their amount, SH and S-S groups in proteins of the yolk vary with fish species. The ionization coefficients value for lateral groups of proteins soluble in the phosphate buffer at 245 and 300 nm is lower for the eggs of the roach than for the eggs of the bream. The Black Sea roach and carp-bream eggs of different degrees of maturity differ from each other by the content of SH and S-S groups of proteins and nonprotein low M_r thiol compounds [*Konovalov*, 1979]. *Fletcher* et al. [1981] have made a characterization of C-reactive protein from the eggs of the marine teleost (*Cyclopterus lumpus*), which is homologous with mammalian C-reactive protein. Its amino acid composition is also described.

Suyama and Suzuki [1978] made a study of free and conjugated amino acids in the extract of rainbow trout (*Salmo gairdneri*) eggs. Among the free amino acids, high levels of aspartic acid, glutamic acid and glutamine are detected, and the sum of the N in these constituents accounts for nearly 50% of the total free amino acid N.Conjugated amino acid content in the extract of eggs is obtained from the difference between the analytical values before and after hydrolysis of the extract, and somewhat larger quantities of serine, glutamic acid, aspartic acid and arginine are determined. The amount of serine N accounts for nearly 30% of the total conjugated amino acid N. In the chromatogram, the presence of a large unknown peak with a shoulder is seen in front of taurine, and the major constituent (X component) emerging in the peak appears to be phosphopeptide, since higher yields of serine and phosphorus are detected along with lower yields of aspartic acid, glutamic acid and arginine in the acid hydrolysate. Alanine glutamic acid and leucine are the most abundant amino acids in the vitellogenin of *S. gairdneri [Hara and Hirai*, 1978; *Cambell and Idler*, 1980], C. auratus [*de Vlaming* et al., 1980] and Japanese eel, *Anguilla japonica* [*Hara* et al., 1980]. *Suzuki and Suyama* [1979] reported the properties of phosphopeptides in the extract of rainbow trout eggs, which contain large quantities of serine and P. The peptides amount to approximately 1% of its eggs and contain more than one half of both the N and P of the extract. Other chemical analyses of vitellogenins in teleosts have demonstrated lower calcium and protein phosphorus contents than those found in vitellogenins of *Xenopus* [*de Vlaming* et al., 1980; *Hara* et al., 1980]. According to *Kim and Zhukinskii* [1978], the amino acid composition of bream (*Abramis brama*) spawn in the process of maturation (from stage IV–V) does not change substantially while conspicuous changes occur during ovum maturation. *Voss* et al. [1978] made a study of isolation, purification and partial characterization of a lectin from chinook salmon (*Oncorhynchus tshawytscha*) ova. Several serological and biochemical studies have demonstrated that the ovaries of many species of the teleost fish families Salmonidae, Percidae, Cyprinidae, Siluridae, Cobitidae and Clupeidae contain powerful agglutinins in various cells [for references, see *Krajhanzl* et al., 1984a, b]. These lectins which are localized in cortical alveoli show expressive serological and sugar-binding characteristics. Several have been purified by affinity chromatography and partially characterized [for references, see *Krajhanzl* et al., 1984b].

Glycoproteins. The carbohydrates and iron occur in the teleost vitellogenins as shown by histochemical techniques [*Hara and Hirai,* 1978; *Hori* et al., 1979; *Hara* et al., 1980]. The iron-binding capacity of vitellogenins in *S. gairdneri,* chum salmon (*Oncorhynchus keta*) and *A. japonica* has been shown with immunological and autoradiographic techniques [*Hara,* 1976; *Hara* et al., 1980]. In addition to the major egg yolk constituents, some other materials are observed. The origin of these substances is not known to be vitellogenin and thus may represent products of endogenous vitellogenesis or exogenous substance of a nonvitellogenin source. *Inoue and Iwasaki* [1978] demonstrated the occurrence of disialosyl(α-N-glycolylneuraminyl-(28)-N-glycolylneuraminyl groups in the carbohydrate chains of a glycoprotein newly isolated from the eggs of rainbow trout *(S. irideus).* Almost all sialic acid residues from disialosyl sequences at the terminal position of the oligosaccharide units consisting of galactose and N-acetylgalactosamine through which the units are linked to threronine residues of the peptide moiety. The sialic acid content of the glycoprotein is as high as 50%. Its unique sugar and amino acid composition is different from that of any class of the glycoproteins so far reported. *Inoue and Iwasaki* [1980] isolated and characterized a family of previously undescribed sialoglycoproteins from the eggs of pacific herring *(C. pallasii)* in a yield 10 times as much as phosvitin which forms the major protein component of the eggs as already discussed. The sialoglycoproteins are composed of 50–60% of carbohydrates and 40–50% of protein. The carbohydrate composition is similar to that of glycoproteins known to have complex types of heterosaccharide units whereas the amino acid composition is unique. The molecular weights ranging over 8,800–13,000 have been determined for the sialoglycoprotein fractions by sedimentation equilibrium. The function of sialoglycoproteins is not known. These may form components of cortical alveoli and chorion (zona pellucida) which show such sialoglycoproteins [*Tesoriero,* 1977b, 1978; *Shackley and King,* 1977; *Wallace and Selman,* 1980] (see also Section A of this Chapter, and Section B of Chapter V).

Carotenoids. The presence of various carotenoids such as canthaxanthin, and lutein-epoxide 4-hydroxy-4-keto-B-carotene, tunaxanthin, astaxanthin ester and asterin acid, has been demonstrated in the eggs of *Oncorhynchus keta* [*Czeczuga,* 1979]. Astaxanthin ester comprises the greatest part, 67.8% of all carotenoids. Various carotenoids such as astaxanthin diester, tunaxanthin monoester, astaxanthin monoester, tunaxanthin, astaxanthin doradexanthin, lutein, zeaxanthin, idoxanthin, triol and tetrol, have also been identified from the ovaries of 9 species of marine fish (Sarodinops) [*Miki* et al., 1982].

Lipids. Ovarian lipids derive from the vitellogenin which appears to form the major source of polar lipids. Total lipids constitute 21.5% of vitellogenin in *S. gairdneri* [*Cambell and Idler,* 1980] and 21–22% of vitellogenin in goldfish *(Carassius auratus)* [*Hori* et al., 1979]. In *S. gairdneri* the vitellogenin separated from other lipoproteins of the serum by centrifugation shows a distinct fatty acid complement from that of the lipoproteins [*Leger* et al., 1981]. Another preparation of vitellogenin from *S. gairdneri* is observed to contain 280 μg polar lipids and 60 μg triglycerides/mg protein [*Wiegand and Idler,* 1982]. Approximately 87% of vitellogenin in *C. auratus* is phospholipid [*Hori* et al., 1979]. A chemical analysis of yolk granules has shown that total lipid constitutes approximately 18% of the dry weight of vitellogenin in the lesser spotted dogfish, *Scyliorhinus canicula* [*Craik,* 1978a]. Of the ovarian derivatives of fish vitellogenin, only lipovitellin contains lipids with amounts ranging from 15 to 29% [*Ando,* 1965; *Jared and Wallace,* 1968; *Plack* et al., 1971; *Cambell and Idler,* 1980]. Employing a conversion factor of 25 X lipid phosphorus determinations the phospholipid content is estimated at 58% of lipovitellin lipid in coho salmon *(Oncorhynchus kisutch)* [*Markert and Vanstone,* 1971] and 60% in cod *(Cadus morhua)* [*Plack* et al., 1971]. *Lizenko* et al. [1979] made a biochemical analysis of lipid contents in the eggs of *Coregonus albula* during the prespawning period. According to their variability, the biochemical indices are divided into three groups: (1) relatively stable lipids (coefficient (CF) of 5–11% – phospholipids, lecithin, cephalin); (2) average stable lipids (CV = 12–30% – total lipids, cholesterol, cardiolipin, sphingomyelin, triglycerides, cholesterol esters, lysolecithin), and (3) unstable lipids (CV = 31–85% – triglycerides, cholesterol esters, lysolecithin; an unidentified phospholipid, cholesterol and caridolipin). Lipid concentration (especially triglycerides and cholesterol esters) in the eggs is affected by the food base of the H_2O body and food ration of the fish.

The accumulation of wax ester has been demonstrated in the ovaries of the opaline gourami (*Trichogaster cosby*). Wax ester forms the major ovarian lipid in this fish, constituting about 37% of ovarian weight [*Sand* et al., 1969]. It also forms the major ovarian lipid in several other teleosts but constitutes a lower proportion of total ovarian mass [*Iyenger and Schlenk,* 1967; *Kaitaranta and Ackman,* 1981]. The source of the extremely high wax ester content in the ovaries of *T. cosby* is of great interest in the overall view of teleost vitellogenesis. Of special interest is the storage of the alcohol portion of wax esters. Its ovarian wax ester alcohols can be synthesized in vivo from dietary acetate, or can derive from dietary carbon chains [*Sand* et al., 1969, 1971]. Its site of synthesis is not known. However, *Wiegand* [1982] has suggested that some ovarian wax ester components may be synthesized endogenously. The plasma fatty acid forms the major potential exogenous source of ovarian alcohols. It is reduced to alcohol after its uptake by the ovary. Since feeding experiments have shown that labelled fatty acid is transported primarily as triglyceride [*Sand* et al., 1973; *Rahn* et al., 1973] there is a possibility that either vitellogenin in *T. cosby* is unique in terms of both its total and neutral lipid content, or other lipoproteins besides vitellogenin facilitate the transport of lipid to the developing ovary. There is a further need for the comparative study of vitellogenesis in fish, as in eggs of viviparous teleosts, not much yolk is formed. Here follicular fluid is present. *Veith* [1979] made a study of osmolarity of the follicular fluid and maternal plasma as well as of analyses of glucose, total sugars, total proteins, total lipids and amino acids in *Clinus superciliosus.* The mean osmolarity of the follicular fluid is 52.48% mosm (milliosmoles)/kg higher than the osmolarity of the plasma (324.2 $\pm$ 20.6 mosm/kg) so that the embryos are always immersed in a liquid medium due to the osmotic influx of water into the follicle. The embryotroph is rich in lipids (76.2 $\pm$ 14.0 mg/100 ml) and amino acids, with the amino acid concentration exceeding normal values for human.

V. Follicular Wall

Follicular wall consisting of zona pellucida, follicular epithelium, basement membrane or basal lamina and theca is developed around the growing oocytes in the fish ovary (fig. 42b, 48, 49) [*Guraya,* 1978]. In their development, structure and composition, these components of the follicular wall show a great diversity in different groups of fishes, especially in elasmobranchs and teleosts which have been extensively used for this purpose [*Kapoor,* 1976b; *Riehl and Schulte,* 1977a; *Riehl,* 1977d, 1978b–e; *Guraya,* 1978; *Emel'Yanova,* 1979; *Donato* et al., 1980]. Especially the zona pellucida (or chorion) shows many variations in its development and structure in oviparous, ovoviviparous, and viviparous fishes. *Makeeva and Mikodina* [1977] also reported several variations in the structure and chemistry of zona pellucida in several cyprinidae *(Hypophthalmichthys molitrix, Aristichthys mobilis, Ctenopharyngodon idella, Mylopharyngodon piceus, Opsaruchthys uncirostris, Pseudorasbora parya, Rhodeus ocellatus, Cyprinus carpio, Hemiculter eigenmanni, Rutilus rutilus, Gobio gobio, Pseudogobio rivularis)* differing in their reproductive characteristics. The covering enveloping the oocytes of *Dicentrarchus labrax,* show numerical, volumetric and structural variations related to the different stages of oogenesis. Up to 6 acellular layers, some of which show a fibrillar disposition and 2 cellular layers corresponding to the follicular and thecal cells, are observed. The developing oocytes in the perch *(Perca fluviatilis)* are larger than the resting oocytes and possess the follicular epithelium which permits the formation of zona pellucida and yolk [*Treasurer and Holiday,* 1981].

A. Follicular Epithelium

1. Morphology and Histochemistry

Initially, a few follicle cells are present close to the very young oocytes of fish ovaries [*Jollie and Jollie,* 1964; *Chaudhry,* 1956; *Anderson,* 1967; *Flügel,* 1967a; *Rastogi,* 1970; *Guraya* et al., 1975; *Shackley and King,* 1977;

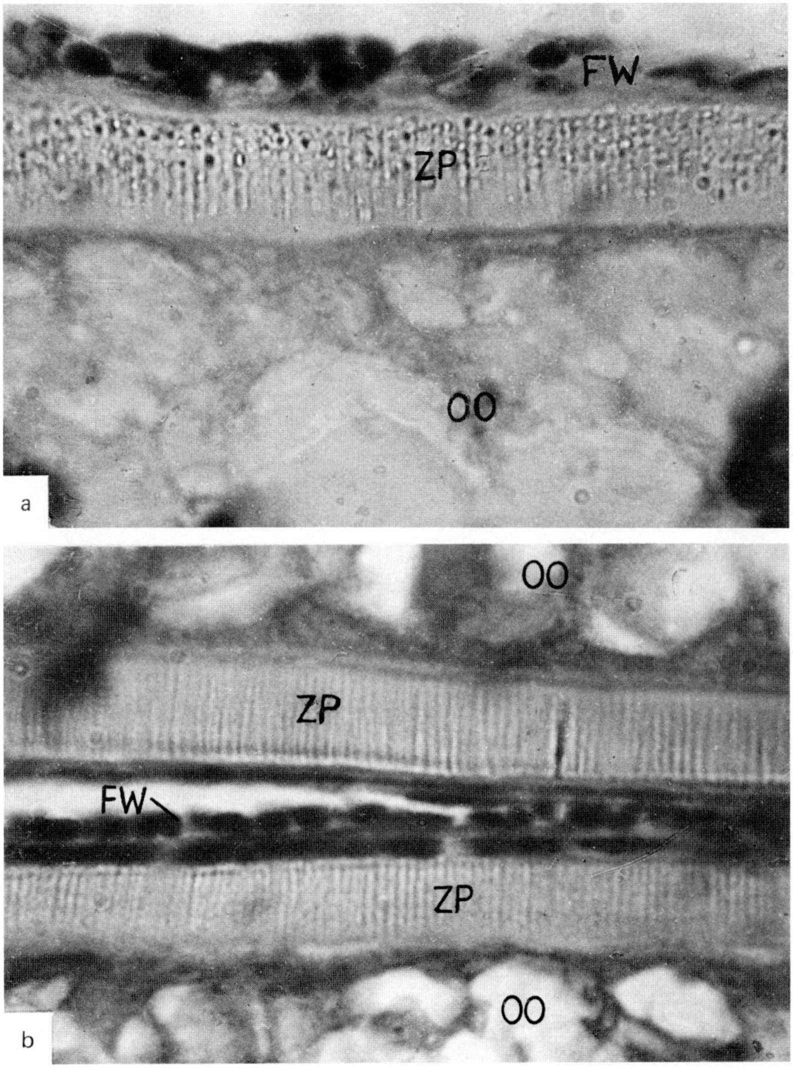

Fig. 48. Portions of large vitellogenic follicles from the ovary of *Cyprinus carpio,* showing a full development of zona pellucida (ZP) having striations and different zones. Very thin follicular and thecal layers constituting the follicle wall (FW) are also seen. *a* Spherical structures of variable sizes, which are usually associated with the striations of the zona pellucida (ZP). Ooplasm (OO) lying adjacent to the zona pellucida (ZP) shows clear cortical alveoli of large size. *b* Striated zonae pellucidae of two closely placed follicles. Eosine-haematoxylin preparation. × 1,000.

Guraya, 1978; *Duslé,* 1980], or a single squamous follicle cell is present as in the case of *Cynolebias melanotaenia* [*Wourms,* 1976]. As the oocyte grows, the follicle cells increase in number, probably by mitosis to constitute a continuous follicular epithelium which retains its single-layered structure throughout oocyte growth in teleosts (fig. 34, 42b, 48, 49) [*Kraft and Peters,* 1963; *Guraya,* 1965, 1976, 1978, 1979a; *Hurley and Fisher,* 1966; *Flügel,* 1967a, b; *Anderson,* 1967; *Götting,* 1967, 1970, 1974, 1976; *Erhardt and Götting,* 1970; *Nicholls and Maple,* 1972; *Gabaeva and Ermolina,* 1972; *Chinareva,* 1973; *Busson-Mabillot,* 1973; *Azevedo,* 1974; *Wourms,* 1976; *Kapoor,* 1976b; *Shackley and King,* 1977; *Tesoriero,* 1977a; *Riehl,* 1977d, 1978b–d; *Riehl and Schulte,* 1977a; *Gabaeva,* 1977; *Emel'Yanova,* 1979; *Toshimori and Yasuzumi,* 1979a, b; *Selman and Wallace,* 1982; *Guraya and Kaur,* 1979, 1982] although a follicular epithelium two cell layers thick has been described in *Fundulus heteroclitus* [*Anderson,* 1966]. In follicles of some fish species (e.g. *Tilapia thollori, Arius thalassinus,* dogfish *Scoliodon sorrakowah*), the follicular epithelium may also show a pseudostratified appearance for some stages of oocyte growth because of the placement of its nuclei in three to four layers [*Kraft and Peters,* 1963; *Gabaeva and Ermolina,* 1972; *Guraya,* 1978]. But this pseudostratified structure is of transitory nature as the follicular epithelium again becomes a monolayer in the advanced stages of follicle growth, suggesting that the pseudostratified condition may reflect cellular reserve to be used during the rapid growth of the oocyte.

With oocyte growth, the follicle cells undergo changes in their morphology. In the beginning, they are usually spindle-shaped or flattened parallel to the oocyte surface or squamous, and have little cytoplasmic differentiation (fig. 50, 51) but just prior to, or at the beginning of vitellogenesis, they gradually show changes in shape from squamous to cuboidal or columnar (fig. 34, 51). As the follicular epithelium is stretched, the amount of cytoplasm in the follicle cells is decreased. The shape of the follicle cell nucleus is correspondingly altered from oblong and ellipsoid to round during follicle growth (fig. 34). The nuclei with prominent nucleoli are placed towards the basal region. In some elasmobranchs the follicle cells show a uniform size, but in others two cell types are differentiated, a large cell with a reticular nucleus and abundant cytoplasm, and a small columnar cell with a densely staining nucleus [*Chieffi,* 1961; *Guraya,* 1978].

In the later stages of oocyte growth, the follicular epithelium is usually stretched, forming some wide intercellular spaces (fig. 49) except at points where the cells are connected by attachment [*Flügel,* 1967a; *Azevedo,* 1974;

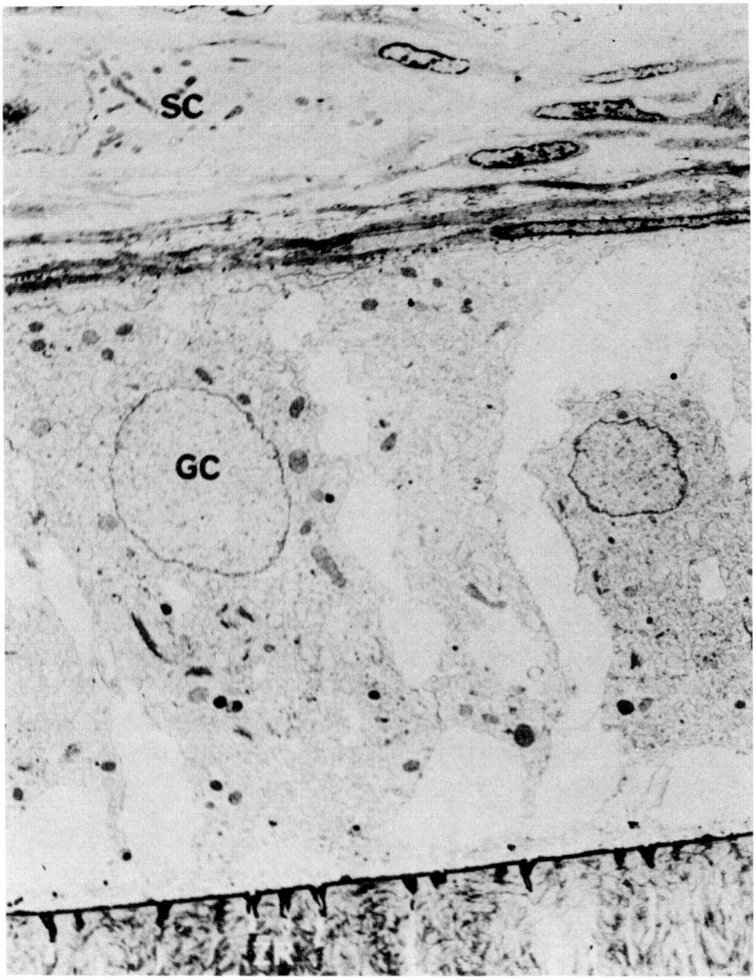

Fig. 49. Survey electron micrograph of a preovulatory follicle of a mature Coho. Note granulosa cells (GC) with dilated granular endoplasmic reticulum containing amorphous materials. Special thecal cells (SC) are seen in the thecal layers. ZR = zona pellucida. × 3,750 [from *Nagahama* et al., 1978].

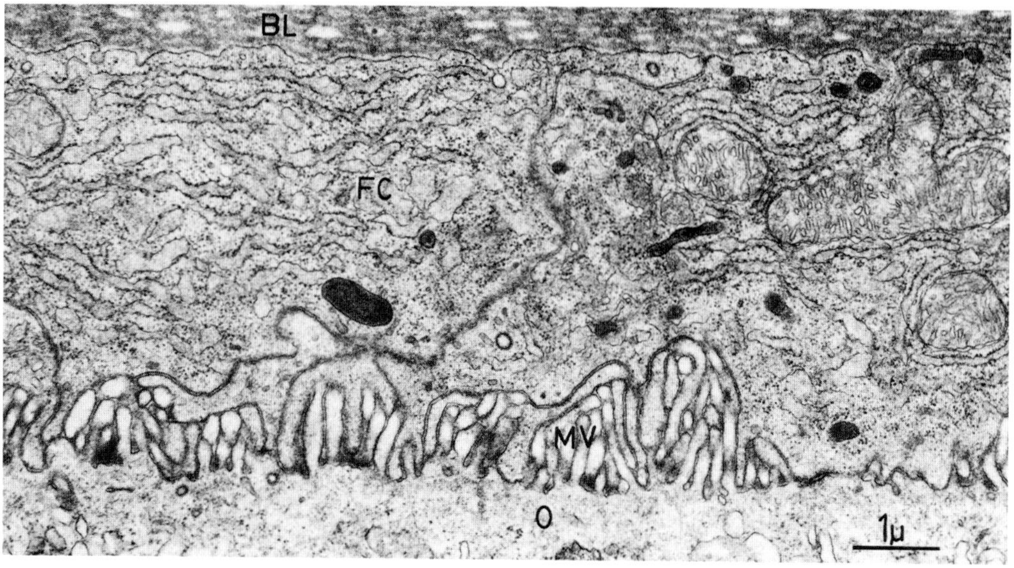

Fig. 50. Electron micrograph of portion of early developing follicle from the ovary of juvenile rainbow trout treated with s-GTH (0.1 μg/g body weight). Follicle cells (FC) surrounded by the basement laminal (BL) show abundant granular endoplasmic reticulum, mitochondria and electron-dense bodies of variable size. Note numerous microvilli (MV) projecting from the oocyte (O) surface [courtesy of Dr. *S.N. Upadhyay*].

Nagahama et al., 1978; *Kagawa* et al., 1981]; the spaces are generally filled with amorphous material, often forming a network-like structure [*Nagahama* et al., 1978; *Kagawa* et al., 1981]. The detailed chemical nature of this intercellular matrix needs to be investigated during follicle growth. The intercellular space between follicle cells in the teleost *Plecoglossus altivelis* is wide except for two focal membrane appositions near the basal lamina [*Toshimori and Yasuzumi,* 1979a, b]. The space between the follicle cells is greatly expanded near the free surface of the epithelium (fig. 49) where it appears to open directly towards the chorion or zona pellucida. Intercellular junctions in the form of desmosomes occur in the follicular epithelium of different teleosts (fig. 51) [*M. Yamamoto,* 1963; *Flügel,* 1964; *Anderson,* 1967]. *Riehl* [1977d] observed that the cells of the follicular epithelium in *Noemacheilus barbatulus* are toothed and contain few desmosomes. In *Gobio gobio* the boundaries of the follicular epithelium are straight and thus more desmosomes occur in this fish than in *N. barbatulus. Toshimori and*

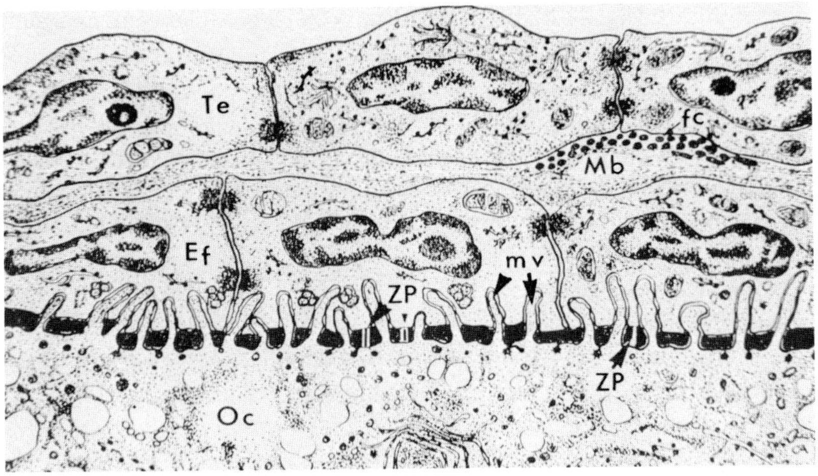

Fig. 51. Diagrammatic representation of the structure of various components of the follicular wall of developing oocyte from *Xiphophorus*. Ef = follicular epithelium; fc = collagen fibres; mv = microvillus; Mb = basement lamina; Oc = oocyte; Te = theca; ZP = zona pellucida [from *Azevedo,* 1974].

Yasuzumi [1979a, b] demonstrated the presence of tight junctions in the follicular epithelium of *Plecoglossus,* which are considered to be 'leaky' or 'intermediate' with regard to the number of strands. Intercellular junctions may be playing some significant roles in the transport of substances through the follicular epithelium.

Gabaeva and Boglazova [1977] studied the morphodynamics and proliferative activity of the follicular epithelium in the oogenesis of *Xiphophorus helleri.* Its maximum proliferative activity corresponds to the period of its morphological transformation. The follicular epithelium of *Hemichromis* during oocyte growth changes from a flattened prismatic into a high prismatic structure [*Gabaeva,* 1977]. Therefore, it can be defined as highly prismatic with secondary specialization. The cross section of the spindle-shaped follicle cells is quadratic in stages I and II and quadrilateral in stage III of oocyte growth in the goby *Pomatoschistus minutus* [*Riehl,* 1978b].

The flattened or squamous follicle cells surrounding the primary oocytes show very little cytoplasmic differentiation which is increased as the follicle cells change in shape from squamous to cuboidal or columnar. Histochemical studies have demonstrated lipid droplets [composed of phospholipids (fig. 42 b), mitochondria, a Golgi complex, and an abundant RNA-containing basophilic substance (or ergastoplasm)] in the follicle cells [*Guraya,* 1965, 1976, 1978, 1979a; *Hurley and Fisher,* 1966; *Ramadan* et al., 1979b; *Guraya and Kaur,* 1982; *Sun and Xizai,* 1983]. The presence of cholesterol in the follicle cells is indicated with polarizing microscopy [*Sun and Xizai,* 1983]. The carbohydrates and proteins are demonstrated besides the various enzyme activities such as LDH, MDH, nonspecific esterase [*Sun and Xizai, 1983].*

Electron microscope studies have revealed the presence of various cell organelles, including elements of granular endoplasmic reticulum, several free ribosomes, a well-developed Golgi complex, smooth-coated vesicles, mitochondria with lamellar cristae, centrioles, dense bodies (secretory or lysosomal), lipid droplets, etc. (fig. 49–51) [*Jollie and Jollie,* 1964; *Hurley and Fisher,* 1966; *Anderson,* 1966, 1967; *Droller and Roth,* 1966; *Flügel,* 1967a; *Götting,* 1967; *Nicholls and Maple,* 1972; *Hirose,* 1972; *Busson-Mabillot,* 1973, 1977; *Azevedo,* 1974; *Chinareva and Kirchinskaja,* 1975; *Wourms,* 1976; *Wourms and Sheldon,* 1976; *Shackley and King,* 1977; *Nagahama* et al., 1976, 1978; *Riehl,* 1977d, 1978b–d; *Riehl and Schulte,* 1977a; *Hoar and Nagahama,* 1978; *Emel'Yanova,* 1979; *Kagawa and Takano,* 1979; *Kagawa* et al., 1981; *Stehr,* 1982; *Ohta and Teranishi,* 1982]. Fine filaments and microtubules are also described in the follicle cells [*Anderson,* 1967]. With the growth of the previtellogenic oocyte, various organelles (especially ribosomes, elements of endoplasmic reticulum, mitochondria, and Golgi complex) progressively increase in amount (fig. 50). The saccules of the Golgi complex acquire a rather dense material. Ultrastructural changes in follicle (or granulosa) cells suggest that these undergo maturation during follicle growth under the influence of hormones (gonadotrophin and possibly steroids). Such hormones have been shown to influence the multiplication and metabolism of granulosa cells of developing follicles in mammals as they develop hormone receptors [*Guraya,* 1985]. It will be rewarding to make similar studies on the follicle cells of developing and maturing follicles of fish ovary as very little work has been carried out previously [*van der Kraak,* 1983].

2. Functions

a. Protein and Lipid Synthesis and Their Transport into the Oocyte

Various electron microscope and histochemical studies have suggested that the follicular epithelium is involved in the synthesis of proteins and lipids during the growth of oocyte. The presence of many ribosomes, ele-

52 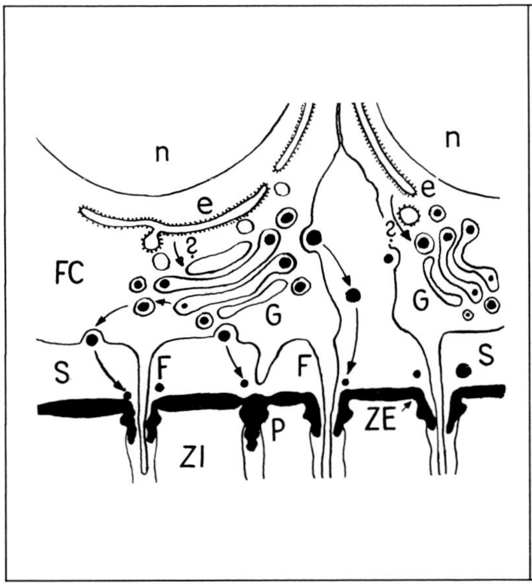

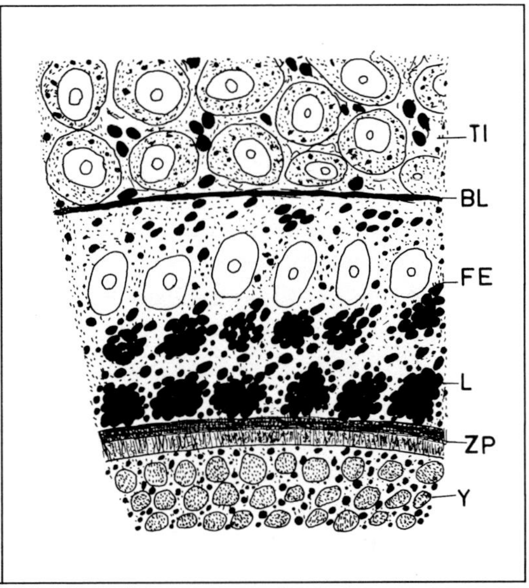 53

Fig. 52. The sites of synthesis, transport and release of material by the follicle cells of *Coregonus albula,* which is possibly used for the formation of plugs in the outer layer (zona radiata externa) of zona pellucida. e = Endoplasmic reticulum; f = follicular epithelium; FC = follicle cell; G = Golgi apparatus; P = plug; S = subfollicular layer; ZI = zona radiata interna; ZE = zona radiata externa [from *Flügel,* 1967a].

Fig. 53. Diagram of portion of vitellogenic follicle from the dogfish *(S. sorrakowah)* ovary showing the distribution of sudanophilic lipids (L) in the hypertrophied theca interna (TI), follicular epithelium (FE) and ooplasm. BL = basal lamina; ZP = zona pellucida; Y = yolk bodies [redrawn from *Guraya,* 1978].

ments of granular endoplasmic reticulum, a well-developed Golgi complex, coated vesicles and dense bodies in the cytoplasm of follicle cells as described above is indicative of protein synthesis (fig. 52). These proteins may be needed partly for the growth and maturation processes of the follicular epithelium itself, and partly for their transport into the oocyte as well as for the formation of zona pellucida (fig. 52). *Chinareva and Kirchinskaja* [1975] demonstrated the synthesis and accumulation of substances in the follicular epithelium of the fish *Coregonus peled* that are released to form the two outer layers of the zona pellucida [see also *Chinareva,* 1975; *Stehr,* 1982]. Similarly, the material of the outer layer (or secondary envelope) of zona pellucida in the growing follicles of the fishes *Cynolebias melanotaenia* and *C. ladigesi* accumulates in the granular endoplasmic reticulum of the follicle cells during stage V [*Wourms,* 1976; *Wourms and Sheldon,* 1976]. It is deposited in stage VI. *Donato* et al. [1980] have observed protein inclusions in the follicle cells of *Chrowis chrowis.* Their material is transported into the oocyte by means of microvilli.

Some phospholipid bodies built up in the follicular epithelium seem to be transported into the ooplasm of growing oocytes [*Guraya,* 1965, 1976, 1978]. Corresponding to the accumulation of yolk, the sudanophilic lipid droplets form large aggregations in follicle cells of the dogfish *Scoliodon ovary* (fig. 53) and meanwhile develop triglycerides in addition to phospholipids [*Guraya,* 1978]. *Lance and Callard* [1969] also reported an increase in lipid droplets during maturation of the follicle wall in the ovary of *Squalus acanthias.* It needs to be determined whether this increase in the lipid droplets of the follicular epithelium in elasmobranchs is indicative of steroidogenesis or simply is related to yolk deposition through providing lipid yolk precursors. The exact mechanisms of synthesis and release of various substances from the follicular epithelium need to be determined more precisely by correlative electron microscope and autoradiographic studies for understanding their precise functional significance.

However, corresponding to the growth of oocyte, the follicle cells develop cytoplasmic processes which penetrate the zona pellucida to a variable distance depending on the stage of oocyte growth as well as on the fish species (fig. 54a–e, 55a–g) [*Jollie and Jollie,* 1964; *Hurley and Fisher,* 1966; *Droller and Roth,* 1966; *Flügel,* 1967a; *Anderson,* 1967; *Götting,* 1967, 1970, 1974, 1976; *Erhardt and Götting,* 1970; *Nicholls and Maple,* 1972; *Busson-Mabillot,* 1973; *Azevedo,* 1974; *Chinareva and Kirchinskaja,* 1975; *Erhardt,* 1976; *Flegler,* 1977; *Riehl,* 1977d, 1978b–d; *Riehl and Schulte,* 1977a; *Stehr,* 1982]. They infrequently interdigitate with the microvilli of the oocyte (fig. 54e, 55b, c). The oocyte microvilli maintain close contact with the follicle cells either by lying next to short microvillar extensions on the surface of the follicle cells within the subfollicular space, or one microvillus from a follicle cell will lie in contact with one microvillus of the oocyte within the pore canal as will be discussed later on. In freeze fracture

Fig. 54. Successive stages (*a–e*) in the development, differentiation and structure of various components of the follicular wall in *S. fontinalis.* Zona radiata externa (ZE) and zona radiata interna (ZI) form zona pellucida which shows follicle cell processes and microvilli lying in its pore canals. In the final stages (*f, g*) of oocyte maturation, the follicle cell processes and microvilli are withdrawn. OP = ooplasm; TF = theca; BM = basement lamina; FE = follicular epithelium [from *Flügel,* 1967a].

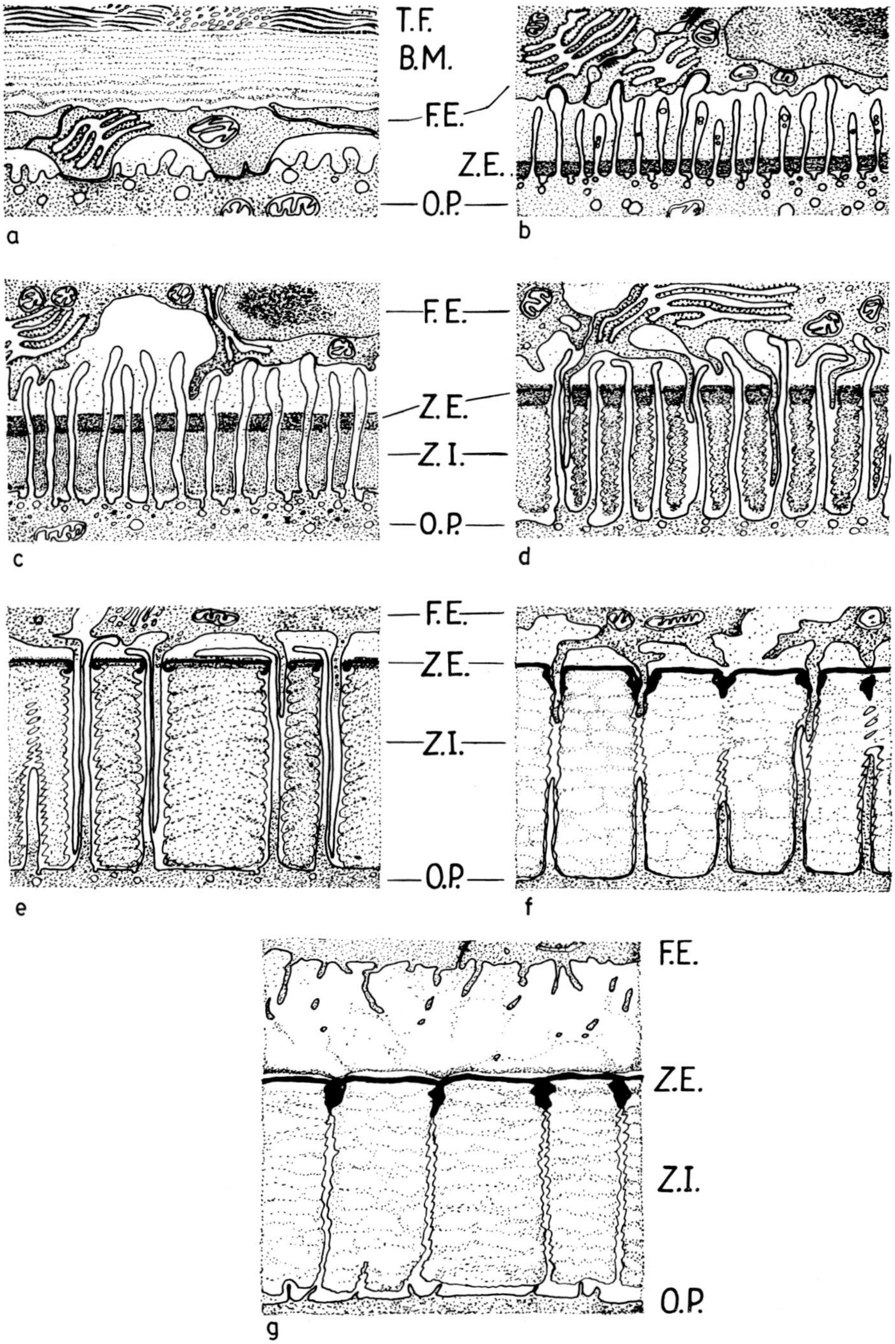
T.F.
B.M.
F.E.
Z.E.
O.P.
a
b
F.E.
Z.E.
Z.I.
O.P.
c
d
F.E.
Z.E.
Z.I.
O.P.
e
f
F.E.
Z.E.
Z.I.
O.P.
g

replicas, many small aggregates of intramembranous particles are demonstrated on the cleavage faces of cytoplasmic membranes of follicle cells. Sometimes, the follicle cell process is closely applied to the oolemma (macula adherens) between the follicle cells and oocytes of *Lebistes reticulatus.* However, *Toshimori and Yasuzumi* [1979a, b] suggested the presence of gap junctions between the oocyte and follicle cells, especially on the surface of follicle cells in the teleost (*Plecoglossus altivelis). Riehl* [1977d, 1978c] observed that from the follicular epithelium shorter and longer processes project into the radial canals of the cortex radiatus (zona pellucida) in the follicles of freshwater teleosts (*Noemacheilus barbatulus* and *Gobio gobio*). In stage III oocytes, the plugs of the cortex radiatus externus (outer layer of the zona pellucida) are attached to the follicular epithelium. The oocyte is anchored to the follicle by the processes of the follicular epithelium [see also *Riehl,* 1976b].

The functions of follicle cell processes and microvilli of the oocyte are specified. *Hurley and Fisher* [1966] suggested the possibility of protoplasmic continuity between follicular cell and oocyte via some microvilli, as evidenced by the absence of intervening membranes between them. This observation needs to be extended and confirmed by using better fixing fluids in electron microscopy, as no cytoplasmic continuity at the points of contact between the microvilli and follicle cell processes is seen in most of the previous studies [*Azevedo,* 1974; *Erhardt,* 1976; *Riehl,* 1977d, 1978b–d; *Stehr,* 1982]. Therefore, the transport of material across their membranes is probably mediated by diffusion and active transport at the molecular level and by pinocytosis [*Hurley and Fisher,* 1966; *Flügel,* 1967a; *Nicholls and Maple,* 1972; *Erhardt,* 1978]. *Azevedo* [1974] observed numerous pinocytotic vesicles in the periphery of oocyte in viviparous teleost (*X. helleri*) (fig. 34, 51), suggesting strong nutritional exchanges between this cell and the surrounding space but, in the present state of knowledge, it cannot be decided about their participation in endocytosis, exocytosis or both. However, *Riehl* [1977d] observed that in early stage I oocytes of *N. barbatulus* and *G. gobio,* there occurs an intense formation of vesicles below the primary oocyte membrane. At the same time, the primary oocyte membrane partially dissolves. In this way, an intimate contact is established between oocyte cytoplasm and follicular epithelium in *N. barbatulus.* This observation needs to be confirmed in future studies. The peripheral ooplasm in *Pomatoschistus minutus* also shows many polysomes besides the considerable pinocytotic activity in stage II [*Riehl,* 1978b]. The possible functions of the pinocytotic vesicles and the polysomes are discussed. Na^+-,K^+-activated

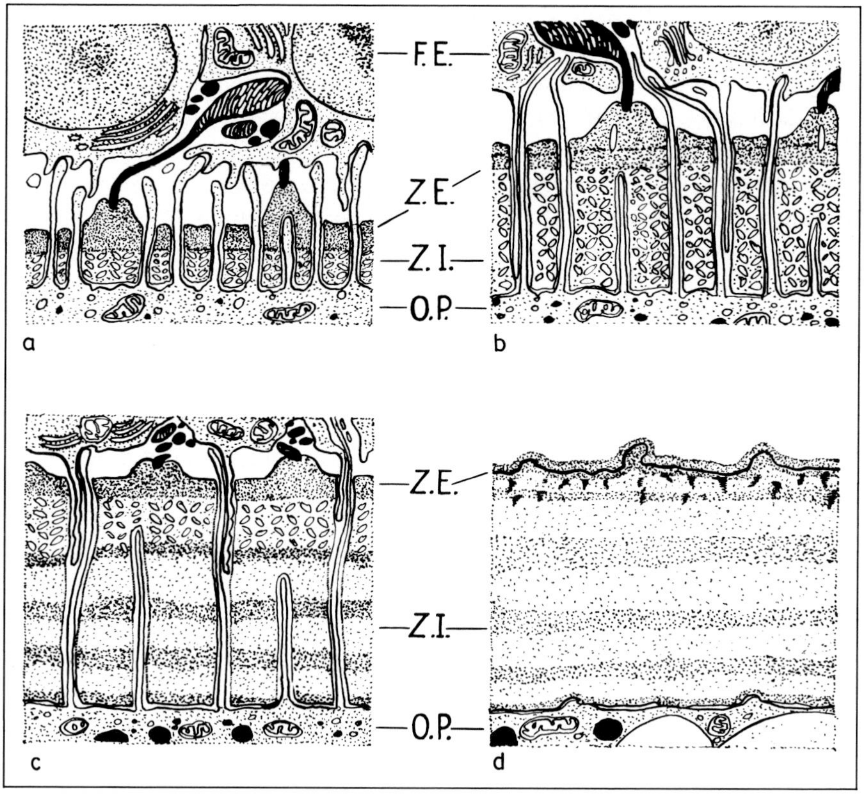

Fig. 55. a–c Successive stages in the development, differentiation and structure of zona pellucida in *F. heteroclitus,* which consists of zona radiata externa (ZE) and zona radiata interna (ZI). Follicular epithelial cells (FE) form processes whereas ooplasmic (OP) surface forms microvilli. Both these components lie in the pore canals of zona pellucida in the mature oocyte. *d* Withdrawal of both follicle cell processes and microvilli after ovulation [from *Flügel,* 1967a].

ATPase observed on the oocyte and follicular microvilli in *Heterandria formosa* [*Riehl,* 1980a] may be involved in the exchange of substances between the follicle and oocyte. *Hurley and Fisher* [1966] attempted to correlate the diameter of the microvilli and the surface area of microvilli with the changing diameter of growing oocytes in the trout *Salvelinus fontinalis.* From these observations they have determined the correlation between morphological alterations of the membranes of follicular wall and the transport of nutrients to the growing oocyte.

The selective transport of substances from outside the follicular epithelium also may occur through the interfollicular cell spaces which contain amorphous material and develop some junctional complexes during follicle growth (fig. 49, 51) [*M. Yamamoto,* 1963; *Flügel,* 1964; *Riehl,* 1977d; *Anderson,* 1967; *Azevedo,* 1974; *Nagahama* et al., 1978; *Toshimori and Yasuzumi,* 1979a, b; *Kagawa* et al., 1981]. The regulation and mechanisms of transport of substances through the follicular epithelium and oocyte surface need to be determined at the molecular levels.

b. Steroid Biosynthesis

The in vitro studies, using elaborate hormone precursors, have shown that ovaries of various fish species are capable of synthesizing a variety of steroids including progestins, corticosteroids, androgens and oestrogens by the Δ^5 (17α-hydroxypregnenolone and dehydroepiandrosterone) or the Δ^4 (progesterone and 17α-hydroxyprogesterone) pathways [*Guraya,* 1976a; *Colombo* et al., 1982; *Fostier* et al., 1983; *Young* et al., 1983b]. Piscine and mammalian gonadotrophins increase the production of ovarian steroids to a variable degree. The regulation of steroidogenesis and steroid activity by various hormonal and external environmental factors has been discussed in detail by *Fostier* et al. [1983]. Species variations in this regard have been revealed. However, the molecular mechanisms involved in the action of gonadotrophin(s) and other hormones for regulating steroid biosynthesis in different ovarian cell types are still to be revealed. *Salmon* et al. [1984] have studied the binding of radioactive iodine-labelled salmon gonadotrophin to theca and granulosa cells from amago salmon ovary. Gonadotrophin receptors have been studied for the fish ovary [*van der Kraak,* 1983]. As in mammals [*Guraya,* 1985], cAMP is also believed to act as the intracellular mediator of gonadotrophin as it relates to studies showing increased cAMP levels or adenyl cyclase activity in ovarian homogenates from several teleosts after incubation with gonadotrophin [*Young* et al., 1983b]; adenylate cyclase has been demonstrated cytochemically in the fish ovary [*Mester* et al., 1980].

Ovarian steroid hormones have been implicated in the regulation of various reproductive processes of fish [*Fostier* et al., 1983]. Although the sites of the production of ovarian steroid hormones in the fish have not been fully established, ovarian follicle layers (follicular epithelium and theca) surrounding oocytes are believed to be the major cellular sites of ovarian steroidogenesis [see reviews by *Guraya,* 1976a, 1978, 1979a; *Hoar and Nagahama,* 1978; *Lambert,* 1978; *Nagahama* et al., 1982; *Nagahama,*

1983; *Fostier* et al., 1983]. Until very recently most of the in vitro studies of steroid biosynthesis dealt with whole ovaries [*Fostier* et al., 1983], but first attempts to isolate cell types of the follicle wall have been reported recently [*Nagahama,* 1983, 1984]. Using electron microscopy and histochemical techniques for the visualization of various hydroxysteroid dehydrogenases involved in the biosynthesis of steroid hormones, very divergent views have been expressed in regard to the role of follicular epithelium in the synthesis of 17β-oestradiol which is well known to be the secretory product of the follicle in the fish ovary [*Nagahama* et al., 1982; *Nagahama,* 1983, 1984; *Fostier* et al., 1983].

In the maturation stage just before ovulation most prominent changes in the follicle cells of teleosts include dilation of vesicular cisternae of rough ER, presence of abundant round, oval or rod-shaped mitochondria with lamellar/tubular cristae and development of lysosome-like bodies with different inner structure and lipid droplets [*Anderson,* 1967; *Yamamoto and Onozato,* 1968; *Nicholls and Maple,* 1972; *Nagahama* et al., 1976, 1978, 1982; *Hoar and Nagahama,* 1978; *Kagawa and Takano,* 1979; *Iwamatsu and Ohta,* 1981; *Kagawa* et al., 1981; *Ohta and Teranishi,* 1982; *Nagahama,* 1983]. The dilated granular endoplasmic reticulum is filled with amorphous material (fig. 56a). The Golgi complex becomes prominent and is generally placed in the apical cytoplasm towards the oocyte surface, it consists of stacks of several flattened cisternae associated with numerous small vesicles. Electron-dense, membrane-bound secretory granules, approximately 100–200 nm in diameter, are frequently seen associated with the Golgi field; some of these granules appear to be in contact with the plasma membrane or in the intercellular space. All these electron microscope studies have suggested that the follicle cells of preovulatory follicles do not show convincing evidence of steroidogenesis as also recently stated by *Ohta and Teranishi* [1982]. Rather they are involved in the synthesis of proteins (fig. 52) as already discussed. However, some workers believe that development of elements of agranular endoplasmic reticulum that are actually intermediate between the smooth and the granular forms and of the mitochondria with tubular or vesicular cristae in the follicle cells of preovulatory and early postovulatory follicles (fig. 56a) in teleosts suggests their possible involvement in steroid production. *Wallace and Selman* [1980] have also demonstrated specific cytological alterations in follicle cells during final oocyte maturation of *Fundulus heteroclitus.* These alterations consist of (1) a proliferation of enormous Golgi complexes with accumulated secretory substance, and (2) an increase in the number of cisternae of granular ER and free ribosomes. These ultrastructural changes are believed to be related to the production of maturation-inducing steroid. This suggestion still needs to be confirmed.

In *Squalus acanthias,* Δ^5-3β-hydroxysteroid dehydrogenase (3β-HSDH) activity is present in the follicle cells [*Lance and Callard,* 1969], which also give a positive reaction for glucose-6-phosphate dehydrogenase (G-6-PDH) [*Lance,* 1968]. Further, both enzyme systems show an increase in their activity in the follicular epithelium of growing oocyte, suggesting that the follicular epithelium may be the possible site for the synthesis or metabolism of steroid hormones. However, contradictory observations have been made in regard to the enzyme activity indicative of steroid hormone synthesis in the follicle cells of teleost ovaries [*Guraya,* 1976a, 1978, 1979a; *Nagahama,* 1983; *Fostier* et al., 1983]. The species differences must be taken into consideration with some reserve as the various studies might not have been carried out at every stage of the ovarian development and maturation. According to *Lambert* [1966], strong 3β-HSDH and G-6-PDH activities are present in the follicle cells of the developing follicle in the guppy *Poecilia reticulata* ovary as also reported for *Tilapia nilotica* [*Sun and Xizai,* 1983]. The G-6-PDH participates in the hexose monophosphate shunt, known to be a necessary process in the synthesis of steroid hormones, and its localization in the follicle cells confirms their steroid-synthesizing activity. In the guppy ovary, 17β-HSDH activity can also be readily demonstrated in the follicle cells, showing that these are most likely the site of oestrogen production in this fish [*Lambert,* 1970a; *Lambert and Van Oordt,* 1974]. The follicular epithelium also shows strong 3β-HSDH activity during vitellogenesis, especially at the exogenous phase in the swordtail, *Xiphophorus helleri* [*Lambert and Van Oordt,* 1974], *Brachydanio rerio* [*Lambert* et al., 1972; *Lambert and Van Oordt,* 1974]; *Sarotherodon aureus, Tilapia aurea* and the mullet, *Mugil capito* [*Livni* et al., 1969; *Livni,* 1971], *Trachurus mediterraneus* [*Bara,* 1974], *Acanthobrana terrae-sanctae* [*Yaron,* 1971], medaka [*Iwasaki,* 1973; *Kagawa and Takano,* 1979] and *Monopterus albus* [*Tang* et al., 1974] and loach [*Ohta and Teranishi,* 1982]. In addition, 17β-HSDH activity has also been shown in the follicle cells of *M. capito* [*Livni* et al., 1969; *Livni,* 1971], guppy [*Lambert,* 1970a], and goldfish [*Khoo,* 1975]. The follicle cells of vitellogenic and ripe follicles in the ovary of *Channa gachua* show 3β-HSDH, 17β-HSDH, 11β-HSDH and G-6-PDH activities, indicating their steroidogenic potential [*Shanbag* and Nadkarni, 1981]. *van den Hurk and Peute* [1979] reported that the follicle cells of *Salmo gairdneri* show a weak but definite 3β-HSDH activity in the

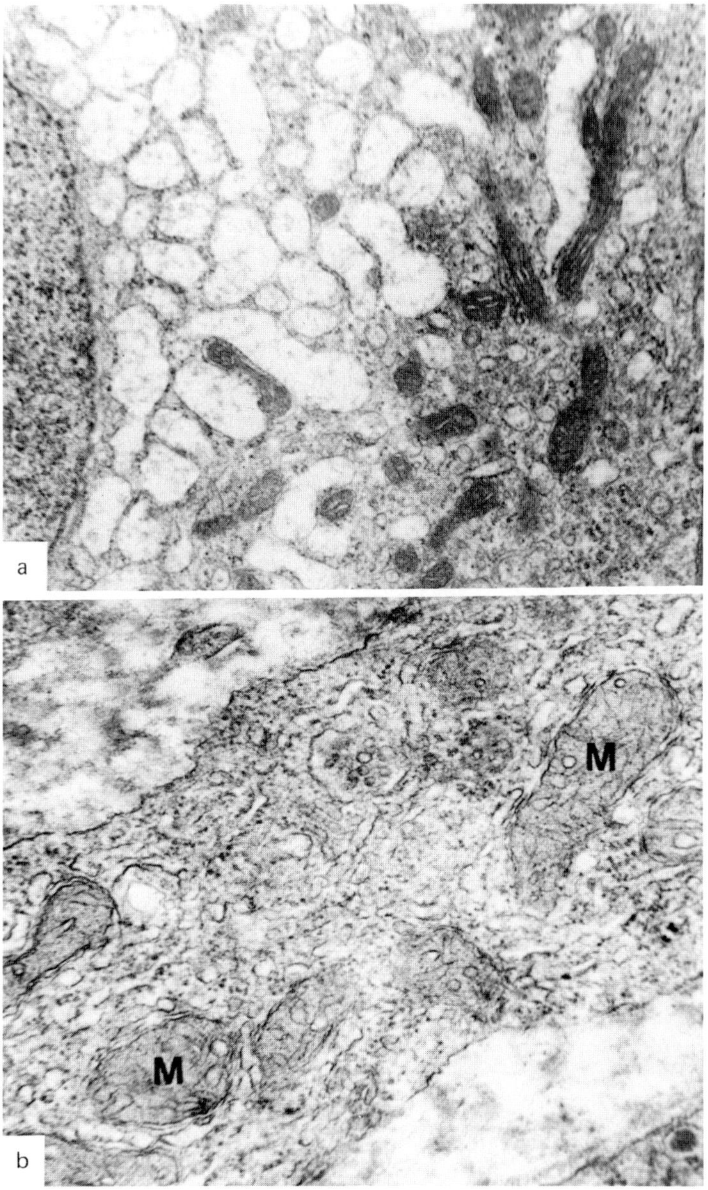

Fig. 56. a Granulosa cells of a mature oocyte from the ovary of the white-spotted char *(Salvelynus leucomaenis).* Note mitochondria with lamellar cristae and dilated granular endoplasmic reticulum filled with amorphous material. *b* Special thecal cell of a preovulatory follicle of a mature oocyte from the ovary of white-spotted char. Note mitochondria (M) with tubular cristae. *a* × 15,400; *b* × 37,100 [from *Kagawa* et al., 1981].

phase of exogenous vitellogenesis. The follicle cells in the ovary of the loach *Misgurnus anguillicaudatus* show a weak but distinct histochemical response for 3β-HSDH only during the periods from prematuration to spawning of the fish [*Ohta and Teranishi,* 1982]. These histochemical studies as well as others [*Iwasaki,* 1973; *Tang* et al., 1974; *Khoo,* 1975] suggest that the follicle cells are likely sources of oestrogens in some teleosts [*Nagahama,* 1983; *Fostier* et al., 1983]. In *Brachydanio rerio, Lambert* [1978] has established the correspondence between the development of the 3β-HSDH activity in the follicular epithelium and the ability of the ovary to secrete 17β-oestradiol and estrone in vitro. The follicle cells are active only during the spawning period in *Oryzias latipes* [*Iwasaki,* 1973], or after ovulation induced with hCG in *Clarias lazera* [*van den Hurk and Richter,* 1980; *Lambert and van den Hurk,* 1982]. However, follicle cells of the preovulatory follicles of *Brachydanio rerio* [*Yamamoto and Onozato,* 1968], *Carassius auratus* [*Nagahama* et al., 1976] and white-spotted char *Salvelinus leucomaenis* [*Kagawa* et al., 1981] do not show 3β-HSDH activity. *Kagawa* et al. [1981] have recently suggested that a lack of correspondence between plasma oestradiol profile and the activity of the follicle cell also does not support the hypothesis that these cells are the principal source of oestradiol in the *S. leucomaenis* ovary. These workers could not correlate any cell type of the follicle with a preovulatory peak of 17β-oestradiol secretion. Further correlative in vitro biochemical and physiological experiments should be carried out to determine the exact steroidogenic role of the follicle cells from the ovaries of different species. However, the recent in vitro biochemical experiments have suggested that both theca and follicular epithelium are needed for the production of 17β-oestradiol as will be discussed in Section E of this chapter.

Lambert [1970a] has observed the presence of 17β- and 3α-HSDH activity in the peripheral ooplasm of older, yolk-laden oocytes, and be suggested that steroids synthesized in the follicular epithelium may also be transported to the oocyte. In this respect, the absence of 3β-HSDH in the oocytes suggests that the latter are not involved in the synthesis of steroid hormones and that the enzymes 3α- and 17β-HSDH occurring at the periphery of yolk-laden oocytes are apparently involved in the intermediary metabolism of steroids secreted by the follicle cells. *Lambert* [1970a] has stated that the physiological meaning of steroid metabolism in the oocytes is difficult to understand. It may be that steroids are required as building material for the oocyte, but it is also possible that they play a role in the disappearance of the germinal vesicle as well as during embryogenesis.

B. Zona pellucida

The acellular layer formed between the surface of oocyte and follicular epithelium (fig. 34, 42b, 48, 49, 54, 55) which will be referred to as the zona pellucida here, shows a great diversity in its structure and chemistry in different groups of fishes. It is also called as the chorion or vitelline membrane or zona radiata or cortex radiatus, etc. [*Guraya,* 1978; *Laale,* 1980]. Synonyms such as coat, covering, pellicle, sac and envelope are also common but provide little information regarding its origin and structural organization [reviewed by *Laale,* 1980]. But it is well established that the materials of zona pellucida are deposited between the oocyte surface and follicular epithelium during growth of follicle, which form microvilli and follicle cell processes respectively (fig. 34, 54, 55). The development, structure and function of follicle cell processes have already been discussed in Section A. Here structural, developmental and functional aspects of microvilli and zona material will be discussed.

1. Microvilli

The plasma membrane of young oocytes lies in close association with the plasma membrane of follicle cells, being separated from it by a narrow space (fig. 34, 54a) [*Jollie and Jollie,* 1964; *Flügel,* 1967a; *Anderson,* 1967; *Erhardt and Götting,* 1970; *Tsukahara,* 1971; *Hirose,* 1972; *Erhardt,* 1976, 1978; *Busson-Mabillot,* 1973; *Azevedo,* 1974; *Riehl and Schulte,* 1977a; *Flegler,* 1966; *Riehl,* 1978b–d; *Upadhyay* et al., 1978; *Emel'Yanova,* 1979; *Stehr,* 1982; *Selman and Wallace,* 1982a, b]. With the oocyte growth, the oolemma develops microvilli, which completely cover the oocyte surface (fig. 34, 50, 54a–c, 55a). Microvilli become long and lie in the space which is formed between the oocyte and follicle cells as the latter move away from the former [see also *Hurley and Fisher,* 1966; *Flegler,* 1977; *Riehl and Schulte,* 1977 a; *Erhardt,* 1978; *Upadhyay* et al., 1978; *Riehl,* 1978b–d; *Stehr,* 1982]. Some of the microvilli are so long that they extend into the intercellular space of the follicle cells (fig. 55b), where they form conspicuous groups but are not branched [*Azevedo,* 1974]. Some microvilli indent the faces of follicle cells lying adjacent to the oocyte (fig. 54b–e, 55a–c). Through gap junctions the microvilli also establish contact with the follicle cells in the teleost *(Plecoglossus altivelis)* [*Toshimori and Yasuzumi,* 1979a, b]. These contact areas appear to be a 7-layered membrane with an overall thickness of 18 μm standard fixation.

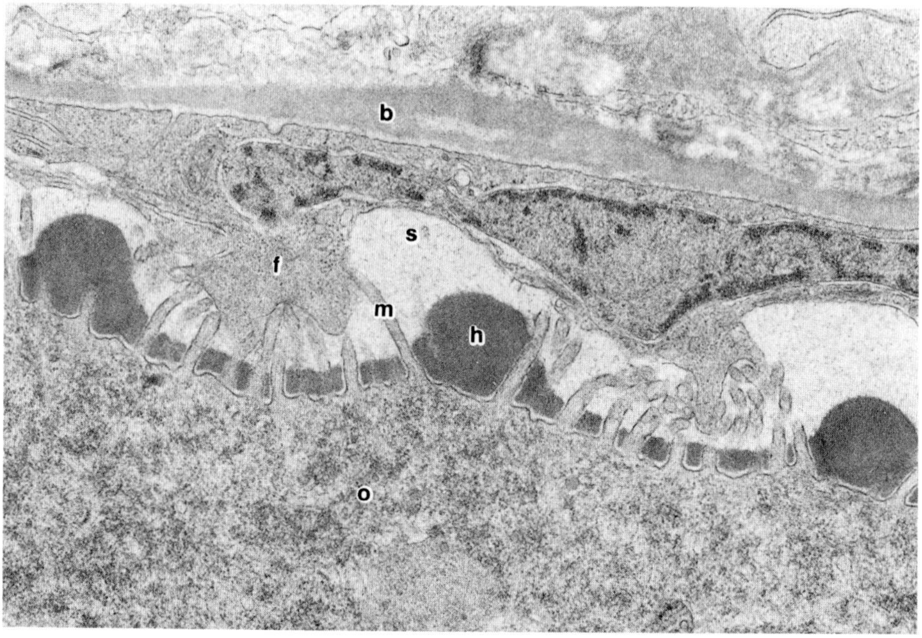

Fig. 57. Portion of stage 3 oocyte of C–O sole *(P. coenosus)*. o = ooplasm; h = hexagon walls of secondary envelope; m = microvilli of oocyte; f = follicle cell; b = basal lamina; s = subfollicular space. × 22,500 [*Stehr and Hawkes,* 1983].

Microvilli of the large or mature oocytes are thin, slender units and have a more dense interior than those of early oocytes [*Jollie and Jollie,* 1964; *Hurley and Fisher,* 1966; *Flügel,* 1967a; *Anderson,* 1967; *Götting,* 1967, 1974, 1976; *Hirose,* 1972; *Busson-Mabillot,* 1973; *Azevedo,* 1974; *Wourms and Sheldon,* 1976; *Wourms,* 1976; *Flegler,* 1977; *Erhardt,* 1976, 1978; *Riehl and Schulte,* 1977a; *Selman and Wallace,* 1982a; *Stehr,* 1982; *Hart and Donovan,* 1983]. These are placed in the pore canals of zona pellucida (fig. 54e–g, 55b, c, 57), as a space is always seen between its components and the microvilli. The substance surrounding the pore canals constitutes the continuous phase of the zona pellucida, whereas the pore canals themselves form the discontinuous phase. However, some very definite striations are observed in the walls of the pore canals [*Hurley and Fisher,* 1966]. These show a 'ribbing' arranged spirally around the pore canals.

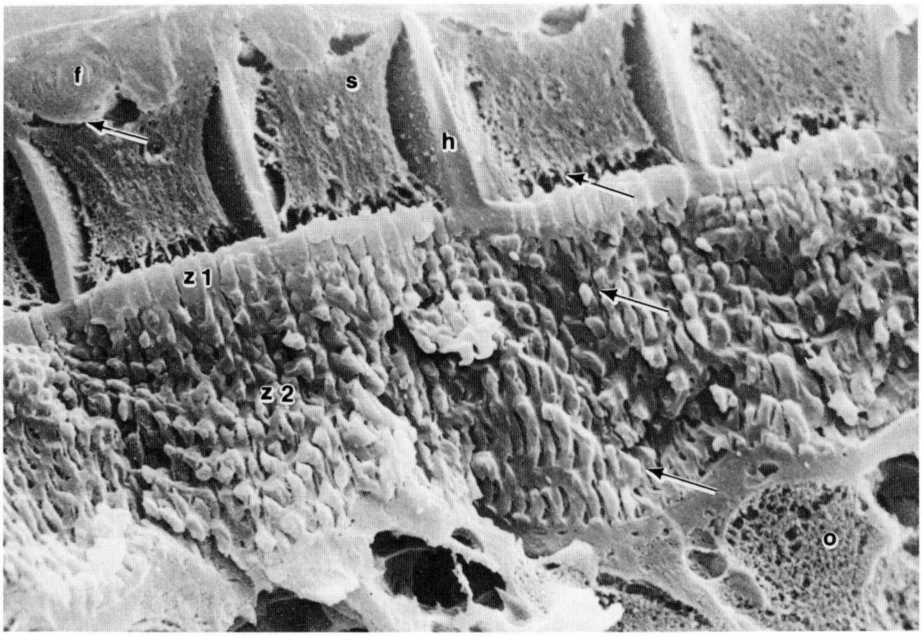

Fig. 58. Scanning electron micrograph of a freeze-cracked, stage 7 oocyte of C–O sole *(P. coenosus)*. Oocyte microvilli can be seen passing through the pore canals across the subfollicular space, and making contact with the follicle cells (arrows). f = follicle cell; h = hexagon walls of secondary envelope; o = oocyte; s = subfollicular space; Z1 = zone 1 of zona pellucida; Z2 = zone 2 of zona pellucida. × 3,000 [*Stehr and Hawkes,* 1983].

The microvilli represent the striations of light microscopic studies [see references in *Kraft and Peters,* 1963; *Guraya,* 1965; *Guraya* et al., 1975, 1977]. As a result of the presence of these numerous striations or microvilli, the layer of zona pellucida lying adjacent to the oocyte has usually been designated as the zona radiata (or cortex radiatus internus) and the outer, relatively homogeneous layer as the zona pellucida proper (or cortex radiatus externus or zona radiata externa) in some earlier studies [*Hurley and Fisher,* 1966, *Flügel,* 1967a; *Riehl,* 1976b, 1978b–d; *Flegler,* 1977; *Riehl and Schulte,* 1977a, 1978; *Erhardt,* 1978]. Microvilli during all stages of oocyte growth show a core of filaments which extend into the microvilli from the ooplasm.

2. Zona Material

a. Structure

During the early stages of oocyte growth, the microvilli do not have an intermicrovillous substance (fig. 34, 50, 54a), which, however, starts to accumulate on the surface of oocyte during further stages of oocyte growth to constitute the zona pellucida (fig. 34, 48, 49, 54b–d, 55a–c, 57, 58, 59) [*Jollie and Jollie,* 1964; *Hurley and Fisher,* 1966; *Droller and Roth,* 1966; *Anderson,* 1966, 1967; *Flügel,* 1967a; *Götting,* 1967; *Erhardt and Götting,* 1970; *Caloianu-Iordachel,* 1971a; *Busson-Mabillot,* 1973; *Azevedo,* 1974; *Chinareva and Krichinskaja,* 1975; *Wourms,* 1976; *Erhardt,* 1976, 1978; *Flegler,* 1977; *Shackley and King,* 1977; *Riehl and Schulte,* 1977a; *Guraya,* 1978; *Stehr,* 1982; *Selman and Wallace,* 1982; *Stehr and Hawkes,* 1983]. Depending on the species as well as on the stage of oocyte growth in the same species, zona pellucida shows variable number of layers (or zones) thus forming either a monopartite, bipartite, or tripartite acellular envelope between the oocyte surface and follicular epithelium (fig. 34, 49, 54b–d, 55a–d, 58, 59) [*Jollie and Jollie,* 1964; *Hurley and Fisher,* 1966; *Droller and Roth,* 1966; *Anderson,* 1967; *Flügel,* 1967a; *Götting,* 1967, 1970, 1974, 1976; *Ginsburg,* 1968; *Erhardt and Götting,* 1970; *Busson-Mabillot,* 1973, 1977; *Azevedo,* 1974; *Chinareva and Kirchinskaja,* 1975; *Erhardt,* 1976, 1978; *Wourms,* 1976; *Wourms and Sheldon,* 1976; *Shackley and King,* 1977; *Riehl and Schulte,* 1977a, 1978; *Riehl,* 1978b–d; *Mikodina and Makeeva,* 1980; *Laale,* 1980; *Stehr,* 1982; *Iino and Inoue,* 1982; *Cherr* et al., 1982; *Hart and Donovan,* 1983]. The major zones can also be distinguished in light microscopic studies (fig. 48) [*Chaudhry,* 1956; *Kraft and Peters,* 1963; *Guraya,* 1965, 1978; *Anderson,* 1967; *Riehl and Schulte,* 1977a] but the details of their structure and formation could not be revealed (fig. 48). In their width, structure and texture, its different zones vary greatly in different teleosts as revealed with electron microscope. According to *Götting* [1967, 1974] the formation of the zona pellucida of oviviviparous and viviparous fishes differs mainly from that of the oviparous species in the decrease of zona layers for making the relationship between the maternal and fetal blood more intimate [see also *Erhardt and Götting,* 1970; *Azevedo,* 1974; *Götting,* 1976].

The tripartite structure of the zona pellucida is generally seen in the oviparous fishes [*Guraya,* 1965; *Hurley and Fisher,* 1966; *Anderson,* 1967; *Götting,* 1967, 1970, 1974; *Busson-Mabillot,* 1973; *Chinareva and Kirchinskaja,* 1975; *Wourms,* 1976; *Erhardt,* 1978; *Treasurer and Holiday,* 1981; *Stehr,* 1982; *Hart and Donovan,* 1983]. *Erhardt* [1978] also recently

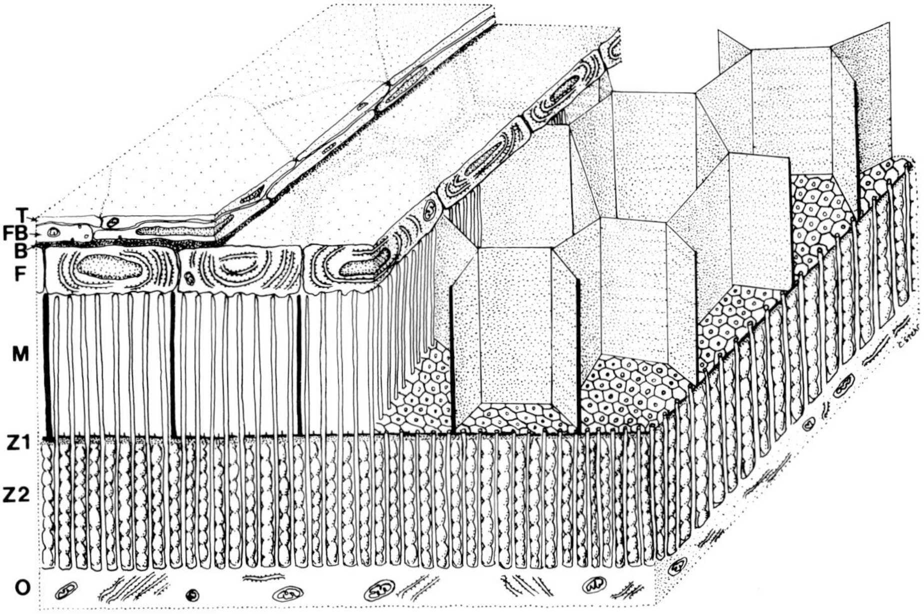

Fig. 59. Schematic drawing of the relationship between the theca cell layer (T), fibroblasts (FB), basal lamina (B), follicle cells (F), hexagonal walls (H), oocyte microvilli (M), polygonal subpattern (arrows), zone 1 of zona pellucida (Z1), zone 2 of the zona pellucida (Z2) and the ooplasm (O) in a stage 7 oocyte of C–O sole *(P. coenosus)*. [*Stehr and Hawkes,* 1983].

observed that the cortex radiatus externus (zona pellucida proper) of the oviparous fish *Lutijanus analis* becomes triple-layered, while the cortex radiatus internus (zona radiata proper) becomes encircled by bundle structures. Three zones, referred to as Z-1, Z-2 and Z-3, have been distinguished in the zona pellucida surrounding the ovarian oocytes of *Hippocampus erectus* and *Syngnathus fuscus* [*Anderson,* 1967] as also reported for *Brachydanio rerio* [*Hart and Donovan,* 1983]. Of these the Z-1 zone reportedly disappears upon the maturation of the egg cell. Zone 1 of the zona pellucida in the oocytes of seahorse *(Hippocampus erectus)* does not develop in the sunfish *(Lepomis microcuris)*, brown trout *(Salmo trutta)*, killifish *(Fundu-*

lus heteroclitus) and other kinds of fish studied by *Fisher* [1963] and *Flügel* [1967b]. It is also not present in the C-O sole oocyte (fig. 59) [*Stehr,* 1982]. The structural organization of zones 2 and 3 in the zona pellucida of the seahorse and pipefish is, nevertheless, the same as that described for the zona pellucida of the teleost *Cynolebias belotti* [*Müller and Sterba,* 1963] and *Agpnus cataphractus* [*Götting,* 1965]. The zones of zona pellucida in the latter two teleosts are also changed in a manner similar to that described for the highly ordered zone 3 in the seahorse and pipefish [*Anderson,* 1967]. The thick zona pellucida of mature oocyte in the white sturgeon *Acipenser transmontanus* consists of four distinct layers [*Cherr* et al., 1982].

Recent electron microscope studies have shown that each layer of zona pellucida may also be composed of as many as three zones [*Anderson,* 1967; *Wourms,* 1976; *Erhardt,* 1978]. Electron microscope studies revealed that the inner layer of zona pellucida in *Crenilabrus melops, Crenilabrus cinereus, Crenilabrus mediterraneus, Centrolabrus exoletus, Ctenolabrus rupestris, Limanda limanda (= Pleuronectes limanda), Hypoglossoides platessoides, Platichthys flesus (Pleuronectes flesus)* and *Gadus morrhua* consists of 6, 7, 9, 17–18, 5–6, 9, 6, 6 and 5 membranous lamellae respectively [*Lönning,* 1972]. Of some interest is the observation by *Lönning* that species of *Pleuronectes platessa* from Bergen and Tromso show variations in ultrastructure, suggesting either a lack of species specificity with regard to zona pellucida organization, or a possible role of the environment in the modification of zona pellucida organization. *Manner* et al. [1977] described the presence of a middle multilayered structure consisting of 19 lamellae in the zona pellucida of *Pimephales promelas.*

The layer next to the oocyte has been described to be a reticular network of lamellae in *Hippocampus erectus* and *Syngnathus fuscus* [*Anderson,* 1967], *Oryzias latipes* [*Hirose,* 1972; *Tesoriero,* 1977a, 1978], *Blennius pholis* [*Shackley and King,* 1977], *Fundulus heteroclitus* [*Kemp and Allen,* 1956; *Flügel,* 1967b]; *Cynolebias melanoteania* [*Wourms,* 1976], *Gobio gobio* and *Noemacheilus barbatulus* [*Riehl,* 1977c], *Pleuronichthys coenosus* [*Stehr,* 1982; *Stehr and Hawkes,* 1983]; complex tubular systems or cross-banded fibrils in *Dermogerys pusillus* [*Sterba and Müller,* 1962; *Müller and Sterba,* 1963; *Flegler,* 1977] and *Cynolebias belotti* [*Götting,* 1965]; lamellar in *Cichlasoma nigrotasrialus* [*Busson-Mabillot,* 1977]; 16 horizontal electron-dense lamellae alternating with 15 interlamellae of lower electron density in *Brahydanio rerio* [*Hart and Donovan,* 1983], and densely packed in *Lebistes reticulatus* [*Jollie and Jollie,* 1964]. The alternating arrangement of obliquely oriented fibrillar lamellae forming ribs occurs in the C-O sole

[*Stehr,* 1982; *Stehr and Hawkes,* 1983] and *Salvelinus fontinalis* [*Hurley and Fisher,* 1966]. *Hurley and Fisher* [1966] suggested that the ribs are arranged spirally around the pore canals. In those teleost eggs with more than one layer, the outer layer has a different ultrastructure from the internal layer, ranging from stratified and fibrous in *Lebistes reticulatus* [*Jollie and Jollie,* 1964], homogeneous in *Blennius pholis* [*Shackley and King,* 1977], and densely packed in *Oryzias latipes* [*Hirose,* 1972].

Stehr and Hawkes [1979] observed that the zona pellucida of starry flounder *(Platichthys stellatus)* constitutes 0.22–0.50% of the egg diameter. It consists of six continuous horizontal lamellae, covered by a thin triple layered border and pierced by numerous regularly spaced pore canals. Thicker membrane of the pink salmon *(Oncorhynchus gorbuscha)* egg forms 0.80–1.0% of the egg's diameter, and is composed of numerous short discontinuous lamellae which are pierced by pore canals and is covered by a coating of irregular thickness. Contrasting environmental conditions during embryogenesis of these two species may be reflected by the thin membrane and simple lamellar structure in the pelagic egg of the starry flounder and the thick membrane complex lamellar structure in the demersal egg of the pink salmon.

Zona pellucida undergoes changes during ovum maturation and ovulation (fig. 54, 55). A rapid size increase is known to occur during maturation and ovulation of teleost oocytes [*Kuo* et al., 1974; *Hirose* et al., 1976; *Hirose,* 1976; *Wallace and Selman,* 1981; *Stehr,* 1982]. Meanwhile, the zona pellucida of *Dermogenys pusillus* show major changes [*Flegler,* 1977]. Zone 1 in the pellucida of seahorse and pipefish, which is first developed, disorganizes during the later stages of oocyte maturation; the significance of its disappearance should be investigated in future studies [*Anderson,* 1967]. Zones 2 and 3 are greatly changed. Similarly, *Busson-Mabillot* [1973] observed that the different zones of the zona pellucida in the teleost *Cichlasoma nigrofasciatum* are greatly changed structurally and chemically during vitellogenesis, preovulation, and ovulation as also reported for the medaka *O. latipes* [*Hirose,* 1972]. When yolk vesicle fusion is complete and the nucleus leaves the centre of the oocyte, the microvilli withdraw from the pore canals. The zona pellucida decreases tine thickness and loses its striated appearance. This type of change also occurs in *Blennius pholis* [*Shackley and King,* 1977], *Fundulus heteroclitus, Pleuronectes platessa* [*Flügel,* 1967a], *Cynolebias melanoteania* [*Wourms,* 1976] and C-O sole [*Stehr,* 1982]. It appears that thinning of the zona pellucida after ovulation is partially a result of stretching the envelope when the egg increases in diameter.

In salmonids, however, the pore canals are plugged with an electron-dense material at the outer surface of the envelope, and the striated appearance of the zona pellucida remains [*Flügel,* 1964, 1967a, 1970; *Hirose,* 1972].

b. Chemistry

Histochemical tests have demonstrated that the zona material in elasmobranchs and teleosts consists of mucopolysaccharides, glycoproteins, carbohydrate-protein matrix or protein and polysaccharide combinations [*Arndt,* 1960a, b; *Stahl and Leray,* 1961; *Guraya,* 1965, 1978; *Anderson,* 1967; *Nakano,* 1969; *Tesoriero,* 1977b; *Makeeva and Mikodina,* 1977; *Pelizaro* et al., 1981; *Lopes* et al., 1982; *Stehr,* 1982]. The biochemical properties of fish egg zona pellucida is different between its individual layers or zones. In the early phases of oogenesis of the flounder (*Liopsetta),* the zona pellucida consists mainly of polysaccharides but later its polysaccharide content diminishes because the protein component predominates [*Nakano,* 1969]. In cyprinoids, however, the zona pellucida contains only protein throughout oogenesis, and a thin hyaline layer outside the zona pellucida contains polysaccharides [*Nakano,* 1969]. In teleosts, such as *Cichlosoma nigrofasciata* [*Busson-Mabillot,* 1977], *Salmo gairdneri* [*Hagenmaier,* 1973] and *Noemacheilus barbatulus* [*Riehl,* 1978c], the inner layer of zona pellucida consists only of protein, whereas its external layer (which is jelly-like in Cichlosoma) contains a combination of proteins and polysaccharides. In *Blennus pholis,* the entire zona pellucida consists only of basic proteins and no polysaccharide material while the external layer also stains for S-H groups, and the internal layer stains for –S-S groups [*Shackley and King,* 1977]. In the C-O sole [*Stehr,* 1982], both zones of zona pellucida show mucopolysaccharides. However, more mucopolysaccharide material appears to be present in the inner layers than in the hexagon walls. *Riehl* [1977c] observed that the cortex radiatus internus (zona radiata) of *Noemacheilus barbatulus* and *Gobio gobio* consists of proteins and neutral lipids whereas the cortex radiatus externus (zona pellucida proper) is mainly composed of polysaccharides but acid mucopolysaccharides predominate in *G. gobio.* The cortex radiatus externus of both species contains some protein. The polysaccharides and acid mucopolysaccharides of the cortex radiatus externus become adhesive after contact with water and in this way the eggs attach to the substrate. The zona radiata (zona pellucida) in lambari *Astyanax bimaculatus* contains neutral polysaccharides and protein radical α-amino groups, cystine, arginine, cystein, tyrosine, and tryptophan [*Lopes* et al., 1982]. *Makeeva and Mikodina* [1977] observed that the zona pellucida (chorion) in different species of Cyprinidae differing in their reproductive characteristics always contains protein and mucopolysaccharides (both acid and neutral). As a rule, the acidic mucopolysaccharides are localized along the outer edge of the zona pellucida or in the ends of villi. In species with a villous zona pellucida, a thin membrane and hyaline layer are always present. *Tesoriero* [1977b] used silver methenamine to reveal ultrastructural localization of polysaccharide material in *Oryzias latipes* oocytes. He reported nonspecific deposition of silver grains, and that the microvilli, attaching filaments, zona pellucida and dense-cored vesicles were most heavily labelled. He also conducted other more specific staining methods which demonstrated polysaccharides on these structures.

The eggs in oviparous fish species have been observed in very diverse aquatic habitats: thus, their zona pellucida has attained a variable structure and chemistry during evolution in order to adapt to the physicochemical characteristics of water. The acidic mucopolysaccharides of the zona pellucida (chorion) are believed to help the eggs to stick to the substrate in Cyprinidae [*Makeeva and Mikodina,* 1977]. Further studies are recommended to investigate the comparative solubility proportion, molecular organization and composition of zona pellucida in different fish species from different water bodies for better understanding the functional roles of zona molecules in the biology of the fish egg. Recent technological advances have allowed the detailed analysis of specific zona pellucida proteins in mammals [*Dunbar and Wolgemuth,* 1984; *Guraya,* 1985]. These advances include the isolation en masse of zona pellucida high-resolution two-dimensional polyacrylamide gel electrophoresis, sensitive protein silver stains, the development of specific monoclonal and polyclonal antibodies; plant lectins have been used in both biochemical and histochemical studies to determine carbohydrate moieties of the glycoproteins present in the mammalian zona pellucida. The application of these advances to the fish zona pellucida will greatly increase our knowledge about the macromolecular composition of its protein and carbohydrate components. The solubility properties of fish zona pellucida can be studied from its sensitivity to both chemical and enzymatic dissolution

c. Origin

The origin of zona material is still controversial. Its synthesis has been attributed either to the oocyte itself or to the follicular epithelium or to both [*Chaudhry,* 1956; *Kemp and Allen,* 1956; *Hurley and Fisher,* 1966; *Anderson,* 1967; *Rastogi,* 1970; *Chinareva and Kirchinskaja,* 1975; *Wourms,*

1976; *Wourms and Seldon,* 1976; *Shackley and King,* 1977; *Tesoriero,* 1977a, 1978; *Sobhana and Nair,* 1977; *Erhardt,* 1978; *Stehr,* 1982; *Lopes* et al., 1982]. The extent to which such a role might differ among species as well as between follicles at different stages of development needs to be investigated in detail. Information regarding the origin of the fish zona pellucida has been obtained mostly by studies using ultrastructural and histochemical methods; very little attempt has been made to apply immunocytochemical methods and autoradiography for this purpose. Although each of these methods are limiting by themselves, their collective application can be of great value for getting an insight into the cellular and developmental regulation of zona pellucida formation during follicle growth.

Anderson [1967] produced an ultrastructural evidence for the synthesis and secretion of zona material by the organelles of the oocyte. During the formation of each zone of the zona pellucida, vesicular and elongate structures originated from the endoplasmic reticulum and Golgi complex are seen in the peripheral ooplasm. The lumina of these structures are confluent with the material constituting the egg coat during its development. *Tesoriero* [1977a, 1978] also suggested that the mechanism of formation of zona pellucida of *Oryzias latipes* apparently involves the transfer of precursor glycoprotein material from the Golgi bodies of the oocyte to the zona pellucida by means of population of dense-cored vesicles. During the early stages of zona pellucida formation in *Cynolebias melanotaenia* oocytes, smooth-surfaced, dense-cored vesicles also fuse with the oolemma and deposit the zona material on the exterior of the oocyte [*Wourms,* 1976]. *Tesoriero* [1977a] distinguished between pinocytotic-coated vesicles and exocytotic smooth-surfaced vesicles as there is a period during yolk synthesis in which yolk is transported into the oocyte by coated vesicles as already discussed in Section B of Chapter IV. C-O sole oocytes at stages 5–7 contain only a few Golgi complexes; however, they show abundant rough endoplasmic reticulum and smooth dense-cored vesicles placed next to the developing striated zona pellucida [*Stehr,* 1982]. This suggests that the oocyte itself may synthesize the zona material of striated layer and transport it in dense-cored vesicles that fuse with the oolemma and release their contents into the surface of the oocyte.

Chinareva and Kirchinskaja [1975] observed that the substances of the outer two layers of zona pellucida, which consists of three layers, are contributed by the follicular epithelium in the fish *Coregonus peled* [see also *Chinareva,* 1975]. *Wourms* [1976] and *Wourms and Sheldon* [1976] also observed that the substance of the outer layer (the secondary envelope) of the zona pellucida in the maturing follicles of annual fishes *Cynolebias melanotaenia* and *C. ladigesi* is synthesized by the follicle cells at stage 5, when they show the ultrastructural features of protein-synthesizing cells. The development of granular endoplasmic reticulum corresponds to the initial development of tubular material. *Shackley and King* [1977] reported changes in the follicular epithelium of marine teleost *(Blennius pholis)*, together with formation of the zona which is considered to be follicular in origin. Two types of follicle cells are described and those appear to function differently in the process of zona formation. The nature of forces regulating all variations in the developmental processes of zona layers of teleosts needs to be determined more precisely [*Laale,* 1980]. The layered structure of zona pellucida suggests that substances having variable physicochemical properties are probably formed at different stages of oocyte growth by the follicle cells and oocyte [*Anderson,* 1967; *Chinareva and Kirchinskaja,* 1975; *Erhardt,* 1976, 1978; *Wourms,* 1976; *Wourms and Sheldon,* 1976; *Riehl,* 1978c; *Stehr,* 1982]. Immunocytochemical studies may reveal biological variation in the cellular origin of zona components among fish species. Resolution of this question may await the application of specific techniques for zona components at the mRNA level. Actually, studies on the synthesis of zona proteins at the transcriptional and translational level are needed for the elucidation of how the macromolecular composition of its unique glycoprotein has diverged whereas the basic physiological functions are highly conserved.

From the review of recent data, as made here, it appears that little progress has been made with reference to problems of origin and naming of egg envelopes. Further correlative ultrastructural, autoradiographic and biochemical studies related to synthesis and secretion, and morphological information on the subcellular organization of both egg and follicle cells are needed to resolve the difficulties still existing. The application of monoclonal antibody techniques will also be very useful to determine the contributions of different cell types in the formation of zona proteins. However, the morphological and chemical differences of the zona pellucida as discussed in different species may reflect adaptations of the fish to diverse ecological conditions. The morphological studies have shown that production of zona pellucida is developmentally regulated as evidenced from its formation during specific stages of oogenesis. But the nature of signals that initiate the zona pellucida formation is not known. These may be hormonal, genetic, or the result of cell contact and establishment of intercellular communication between the oocyte and follicle cells. Whatever the signals may be, the regulation of zona pellucida production could occur at the level of posttranslational modification, translation, transcription, and/or genome organization. The molecular basis of regulation must be determined in future studies with recombinant DNA and monoclonal antibody probes. Since intraspecies, biochemical, histochemical and morphological differences occur in the fish zonae, it is possible that the site of synthesis and the posttranslational modification may also vary among fish species. The cellular origin of zona pellucida components can also be determined using in vitro biosynthesis of proteins and carbohydrates.

C. Surface Structures of Eggs

The morphological features of fish egg envelopes are highly adapted to the environmental conditions in which the embryo develops [*Ivankov and Kurdyayeva,* 1973; *Guraya,* 1978; *Stehr and Hawkes,* 1979; *Laale,* 1980; *Brummett and Dumont,* 1981]. *Laale* [1980] has critically reviewed the earlier observations regarding membrane adhesiveness and modifications in the eggs of different species of teleosts. Demersal eggs, which are often subjected to abrasive forces, generally develop thick envelopes with complex lamellae [*Hurley and Fisher,* 1966; *Flügel,* 1967a; *Ivankov and Kurdyayeva,* 1973; *Osanai,* 1977; *Manner* et al., 1977; *Stehr and Hawkes,* 1979, 1983]. Such eggs may also have elaborate surface structures that perform specialized functions. For example, the eggs of different teleosts possess some adhesive devices for their attachment to the substratum. Fibrils on the surface of *Fundulus heteroclitus* eggs [*Tsukahara,* 1971; *Dumont and Brummett,* 1980; *Brummett and Dumont,* 1981] facilitate adhesion to each other and to the substrate [*Anderson,* 1974] and help to retain moisture when the eggs are exposed at low tide [*Kuchnow and Scott,* 1977]. Hollow spines occur on the surface of *Cynolebias melanotaenia* eggs [*Wourms and Sheldon,* 1976; *Wourms,* 1976; *Sponaugle,* 1980], and these may partially protect the eggs when these are exposed to air during the dry season [*Wourms,* 1976]. In contrast to demersal eggs, the pelagic eggs with a thin zona pellucida generally show smooth surfaces. Their thin zona pellucida consists of horizontal lamellae [*Ginsburg,* 1968; *Hagstrom and Lönning,* 1968; *Lönning and Solemdal,* 1972; *Ivankov and Kurdyayeva,* 1973; *Lönning and Hagstrom,* 1975; *Stehr and Hawkes,* 1979, 1983]. A few species of teleosts with pelagic eggs also show elaborate surface structures such as regularly spaced bristles, protuberances, or raised hexagonal patterns [*Ahl-*

strom and Moser, 1980]. However, our knowledge is very meagre about their origin during oogenesis and function in the mature egg [*Robertson,* 1981; *Stehr,* 1982]. The development of teleost egg zona pellucida has been studied extensively for fish with smooth egg surfaces as already discussed. However, development of elaborate surface structures of teleost egg zona pellucida during oogenesis has been studied in only a few species with demersal eggs, which include *Cynolebias melanotaenia* [*Wourms,* 1976; *Wourms and Sheldon,* 1976; *Sponaugle,* 1980], *Cichlasoma nigrofasciatus* [*Busson-Mabillot,* 1977] and *Fundulus heteroclitus* [*Kemp and Allen,* 1956; *Anderson,* 1966]. Our knowledge is poor about the fine structure of the zona pellucida of the fish eggs during and after fertilization. However, *Perry* [1984] using scanning electron microscopy has observed that the unfertilized egg of the winter flounder *Pseudopleuronectes americanus* shows a cross pattern of depressions which radiate in all directions across the zona surface, creating a wrinkled appearance. After fertilization, the surface of zona pellucida becomes regular with a smoother appearance. The pores of the unfertilized egg are flush with the zona surface, but become thickened and elevated after fertilization. The zona pellucida of the fertilized egg becomes granular in contrast to the smooth and uniformly textured zona pellucida of unfertilized egg.

The development, structure and nature of adhesive devices during fish oogenesis form the subject of some recent studies. *Rubstov* [1978] studied the structural characteristics of egg envelope in *Cyprinus carpio* and *Clupea harengus* in the region of mucilage attachment to substrate (vegetation). The envelopes of the two fish differ greatly according to individual structures in spite of the fact that characteristics of spawning grounds, substrates for egg laying and adaptive characteristics of their embryogenesis are very similar. *Busson-Mabillot* [1977] described the development, structure, chemistry and function of adhesive apparatus of a teleost fish's *(Cichlasoma nigrofasciata)* egg. It fixes the eggs to the substratum after their laying. The adhesive apparatus consists of two distinct elements: on the one hand, filaments connected to the zona pellucida surface and on the other hand to a mucous jelly coat which externally covers the zona pellucida. Follicle cells successively secrete these two elements during the course of vitellogenesis. The filaments consist of bundles of protein tubules about 20 nm in diameter. At the start of vitellogenesis, the tubular proteins synthesized by the rough endoplasmic reticulum are discharged directly into the extracellular space where these polymerize into continuous tubules. Later on, the glycoproteins of the jelly coat are synthesized and accumulate as voluminous granules in the rough endoplasmic reticulum of follicle cells. Discharge of these glycoproteins occurs during ovulation and proceeds according to a type of apocrine secretion. *Anderson* [1966] also associated the formation of fibrilla on *Fundulus* eggs with an increase in microtubules, mitochondria, rough endoplasmic reticulum, ribosomes, and Golgi complexes in the follicle cells. *Kemp and Allen* [1956] observed extracellular strands of filaments among *Fundulus* follicle cells and concluded that the fibrils were derived from the filaments and intercellular matrix of the follicle cell processes. The attaching filaments in the oocytes of *Cyprinodon variegatus* and *Oryzias latipes* are of two types having different thickness [*Iino and Inoue,* 1982]. In cross section, these appear to be composed of tubular units.

According to *Riehl* [1978b] up to 220 attaching filaments originate in stage II of oocyte growth in the goby *Pomatoschistus minutus* and are connected with the egg envelope at the animal pole. These filaments form a modified part of the cortex radiatus externus (zona pellucida proper) and consist of two substances which differ in their electron density. The attaching filaments lie between the cells of follicular epithelium and are probably formed by these cells. These are adapted for attaching eggs to the substrate. Scanning electron microscope study of eggs in Acipenseridae has revealed a sticky surface on their gelatinous layer which is capable of special movements enabling it to adhere to a substrate [*Markov,* 1978]. According to *Takahashi* [1978] the adherent filaments in the eggs of isaza *(Chaenogobius isaza)* are formed during the tertiary yolk globule stage between the zona pellucida and follicular epithelium. The filaments show a reticular structure around the micropyle, and run almost parallel to the meridional lines on the wall of the egg surface. As the egg is released into water and sticks to the wall of the vessel, the adherent filaments detach from the egg envelope and turn inside out, except around the micropyle. In this way, the stalk of the egg is formed. In other species of gobiid fish, the structure and origin of adherent filaments are essentially the same. The eggs of osmerids and salangids, though only distantly related to gobiids, also have adherent envelopes (filaments) of similar structure. The elaborate zona pellucida of *Cynolebias melanotaenia* is covered with uniformly spaced, hollow conical projections which terminate as a crown of recurved spikes [*Wourms,* 1976]. These are arranged in equilateral triangles arising from a pentagonal base, which is part of a surface raised in a pattern of pentagon and hexagon interconnected ribs. Small vesicles containing 25-nm tubules form within the rough endoplasmic reticulum and are then transported to the exterior of the cell and assemble into the ornamentation of the zona pellucida

[*Wourms and Sheldon,* 1976; *Sponaugle,* 1980]. The C-O sole *(Pleuronichthys coenosus)* is one of five species within the genus *Pleuronichthys* that have eggs with raised hexagonal surface structures (fig. 59) [*Sumida* et al., 1979]. The species in this genus are unusual in that they are among only a few teleosts in Pacific Northwest waters that have pelagic eggs with ornate surface structures. Ultrastructural studies of C-O sole eggs have revealed that ornate hexagonal surface structures are formed from glycoproteins secreted by the follicle cells [*Stehr,* 1982; *Stehr and Hawkes,* 1983]. They first appear as a thin osmiophilic layer evenly dispersed on the surface of the young oocytes averaging 0.12 nm in diameter (fig. 57). The vertical walls or struts of the hexagons start as discrete thickenings in the osmiophilic layer that become stout extensions and then are lengthened and thinned to a final height averaging 0.09 nm (figs. 58, 59). The oocyte surface is covered by closely-associated follicle cells and the walls of the hexagons are formed in the space between these cells (fig. 57, 58). The hexagonally patterned zona pellucida studied by *Stehr and Hawkes* [1983] in detail most likely provides the egg with additional protection, resiliency and bouyancy during embryogenesis. The functional significance of pelagic egg surface extensions including their possible involvement in gas exchange needs to be determined more precisely in future studies.

In most oviparous fishes, ovulation terminates the relationship of the developing ovum and its follicular attachment. In various fishes, follicular contribution to the external egg layer is by secretion of the follicle cells as already discussed. But *Sheldon* [1978] observed the development and transformation of the follicular epithelium into a functional egg envelope in *Dorosoma petenense.* The highly adhesive egg capsule of *D. petenense* derives from transformed ovarian follicular epithelium. The follicular epithelium of the oocytes develops from a loosely organized layer of cuboidal cells into a simple columnar cell layer, then into a granular layer and finally into a layer with a radially striated structure. This altered membrane is not ruptured at ovulation but remains intact throughout incubation and is the source of eggs' tenacious adhesiveness.

D. Micropyle

The great development and complexity of the zona pellucida appear to be responsible for the development of micropyles during oogenesis in the eggs of different teleost species (fig. 60) [*Ginsburg,* 1968; *Rastogi,* 1970;

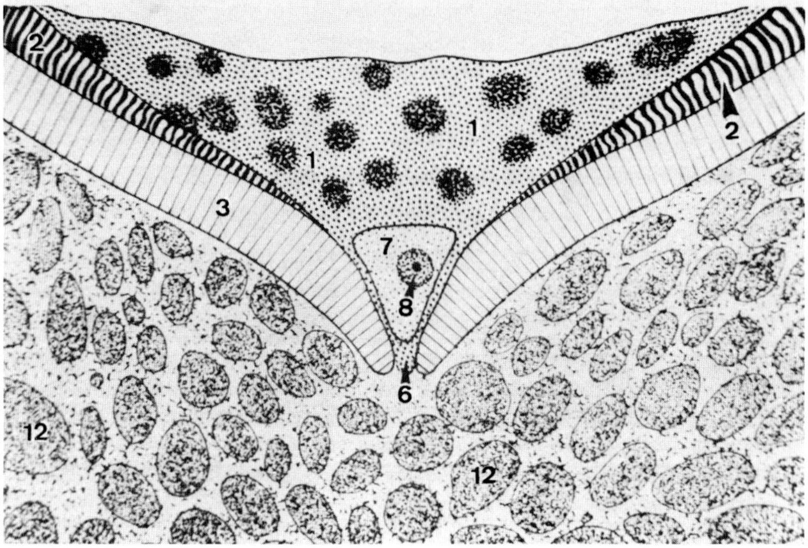

Fig. 60. The development of micropyle in the oocytes of *Gobio gobio.* 1 = follicular epithelium; 2 = external cortex radiata; 3 = internal cortex radiata; 6 = canal of micropyle; 7 = cone cell; 8 = nucleus of cone cells; 12 = yolk [from *Riehl,* 1977a].

Riehl and Götting, 1974, 1975; *Szöllösi and Billard,* 1974; *Kuchnow and Scott,* 1977; *Riehl,* 1977e, 1978d, e, 1979, 1980b; *Riehl and Schulte,* 1977b, 1978; *Stehr and Hawkes,* 1979; *Brummett and Dumont,* 1979; *Hosokawa,* 1979; *Dumont and Brummett,* 1980; *Kudo,* 1980; *Laale,* 1980; *Hosokawa* et al., 1981; *Iwamatsu and Ohta,* 1981; *Cherr* et al., 1982; *Hart and Donovan,* 1983]. *Laale* [1980] has reviewed the earlier light microscope observations regarding the structure, function and origin of micropyle in different species of bony fishes. The structure and development of micropyles, which delineate the animal pole of the oocyte, greatly vary in different teleosts (fig. 60, 61). Actually, the zona pellucida ultrastructure surrounding the distal and proximal micropyle openings, the size of the openings and the shape of the canal differ between species [*Szöllösi and Billard,* 1974; *Riehl and Götting,* 1974, 1975; *Riehl,* 1977e, 1978d, e, 1979; 1980b; *Riehl and Schulte,* 1977b; *Kuchnow and Scott,* 1977; *Osanai,* 1977; *Hosokawa,* 1979; *Stehr and Hawkes,* 1979; *Dumont and Brummett,* 1980; *Stehr,* 1982; *Cherr* et al., 1982; *Hart and Donovan,* 1983]. The formation of micropyle depends on

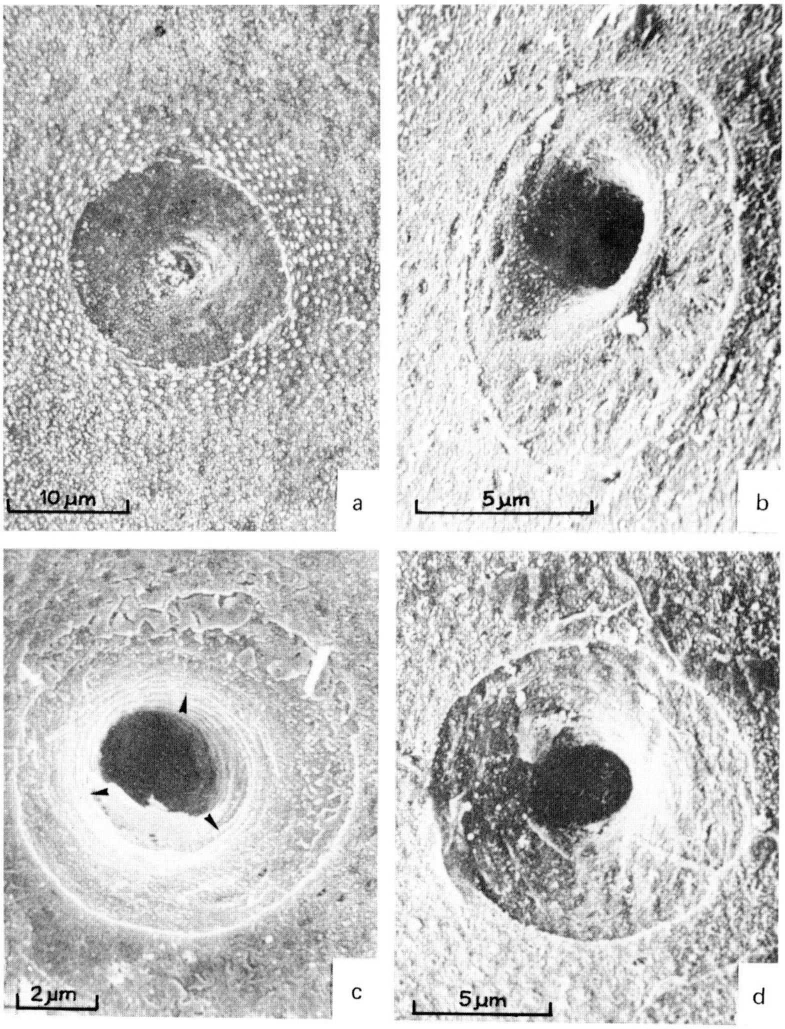

Fig. 61. Micropyles. *a* Micropyle of *Salmo gairdneri.* The micropyle funnel is surrounded by a wreath of little knobs, which are characteristic of this species. *b* Micropyle of *Salmo trutta.* The opening of the micropyle canal is not situated in the centre of the micropyle funnel. *c* Micropyle of *S. trutta* morpha fario. The arrowheads point at reinforcements of the sides of the micropyle canal. *d* Micropyle of *S. trutta* morpha lacustris [from *Riehl,* 1980b].

the disposition of the surrounding layers of the follicular wall. *Riehl* [1977e] observed that the micropyle of *Noemacheilus barbatulus* is established in late stage I and its development is completed in late stage III of oocyte growth. The micropyles of *N. barbatulus* and *Gobio gobio* are formed by a modified follicle cell (fig. 60) which can be seen at the late perinucleolus stage in isaza, *Chaenogobius isaza* [*Takahashi,* 1981]. This modified follicle cell sends out a thin apophysis, which leaves open the micropyle canal. The micropyle of *N. barbatulus* is formed by the cortex radiatus externus (zona radiata), while the micropyle of *G. gobio* is formed only by the cortex radiatus internus. *Osanai* [1977] studied the envelopes surrounding the chum salmon *(Oncorhynchus keto)* oocytes in the preovulatory and postovulatory stages. The follicle cells adhering to the zona pellucida of ovarian oocytes are dispersed at ovulation. The micropyle becomes observable in the ovulated ripe eggs. *Sobhana and Nair* [1977] made some observations on the origin and fate of microplye in the oogenesis of olive carp *(Puntius sarana).* In the oocytes of *N. triangularis,* the micropyle is formed by the coalescence of three enlarged follicle cells [*Kumari and Nair,* 1979]. The micropyle canal in the eggs of starry flounder *(Psettodes stellatus)* measures 8 µm at the opening and tapers to 3.6 µm as it penetrates the membrane [*Stehr and Hawkes,* 1979]. The 15–16 µm micropyle opening in pink salmon *(Oncorhynchus gorbuscha)* egg is surrounded by an area of protrusion and the funnel-shaped canal tapers to 2 µm at its terminal structure. Numerous micropyles (average 7) are restricted to a 100–200 µm region at the animal pole and penetrate the zona pellucida of eggs in the white sturgeon, *Acipenser transmontanus* [*Cherr* et al., 1982]. Their outer opening measures 15 µm in diameter. The micropylar canal tapers twice, ultimately terminating at the oolemma with an inner opening of diameter of 1.2 µm. The micropyles of the white sturgeon egg appear more complex than the micropyles in other fish egg. The micropylar apparatus in the egg of *Brachydanio rerio* consists of a cone-shaped vestibule and a tapered canal traversing the zona pellucida [*Hart and Donovan,* 1983]. The outer diameter of the canal is 7.5–8.5 µm and inner diameter 2.3 µm. The inner micropylar aperture is differentiated as a circular cluster of 10–20 microvilli-like projections. The cortical granules are absent in the cytoplasm directly below the sperm entry site.

Riehl [1978e] studied the problems of development, structure and function of micropyle in the oocytes of the goby *Pomatoschistus minutus.* The micropyle in this species is situated at the animal pole between the points of departure of attaching filaments and consists of a channel perforating the cortex radiatus (zona pellucida). There is no micropyle pit and thus the micropyle of *P. minutus* belongs to type 3 of micropyles described by *Riehl and Götting* [1974, 1975] and *Riehl* [1980b]. This micropyle, the smallest of teleosts yet studied, is formed by a modified cell of the follicle epithelium, the so-called plug cell, which differs from other cells of the follicular epithelium by its shape and size. The eggs of *P. minutus* are adherent to the substrate by the micropyle pole. Problems of insemination caused by this fact are discussed.

The micropyle continues to be occupied by the highly specialized micropylar cell until matured oocytes are released from their enclosing egg envelopes at the time of ovulation. The triangular, large micropylar cells show granular ER, mitochondria, small vesicles, Golgi complex, filaments and microtubules, which show a specific distribution within the cytoplasm [*Ohta and Teranishi,* 1982; *Nagahama,* 1983]. The apical cytoplasm contains numerous microtubules which lie parallel to the long axis of the cell. The cisternae of granular ER containing amorphous material are present in the basal cytoplasm, indicating their secretory function. The micropylar cells do not show ultrastructural signs indicative of steroidogenesis in the ovary of loach *M. anguillicaudatus;* however, these cells show an intense reaction for 3β-HSDH at the yolk globule stage [*Ohta and Teranishi,* 1982]. This intense reaction of the micropylar cells is seen throughout the year, being in sharp contrast to the follicle cells which lose the enzyme activity during the postspawning and sexually quiescent period.

Riehl [1980b] using SEM studied the micropyle of 7 salmonids and 4 coregonids. He suggested that the micropyles and the surface pattern of fish eggs form important characteristic criteria for identification of different teleosts (fig. 61, 62) [see also *Riehl,* 1978e, 1979; *Riehl and Schulte,* 1978]. The micropyle studied belongs to two different types: (1) micropyle with flat pit and long canal (type II) and (2) micropyle with canal but without pit (type III) (fig. 63). *Riehl and Götting* [1974] also reported the third type with a hollow pit leading into a short canal (fig. 63). The sides of the micropyle canal are often reinforced by annular and helicoid thickenings [*Szöllösi and Billard,* 1974; *Riehl,* 1979; *Stehr and Hawkes,* 1979]. These concentric ridges within the micropyle canal probably correspond to edges of the horizontal lamellae of the zona pellucida. SEM studies of the funnel-shaped micropyle in mature eggs have revealed that the wider distal opening leads into a canal which tapers to an aperture just large enough to permit entry of a single sperm [*Kuchnow and Scott,* 1977; *Stehr and Hawkes,* 1979; *Ohta and Teranishi,* 1982]. Since the diameter of the inner micropylar aperture in

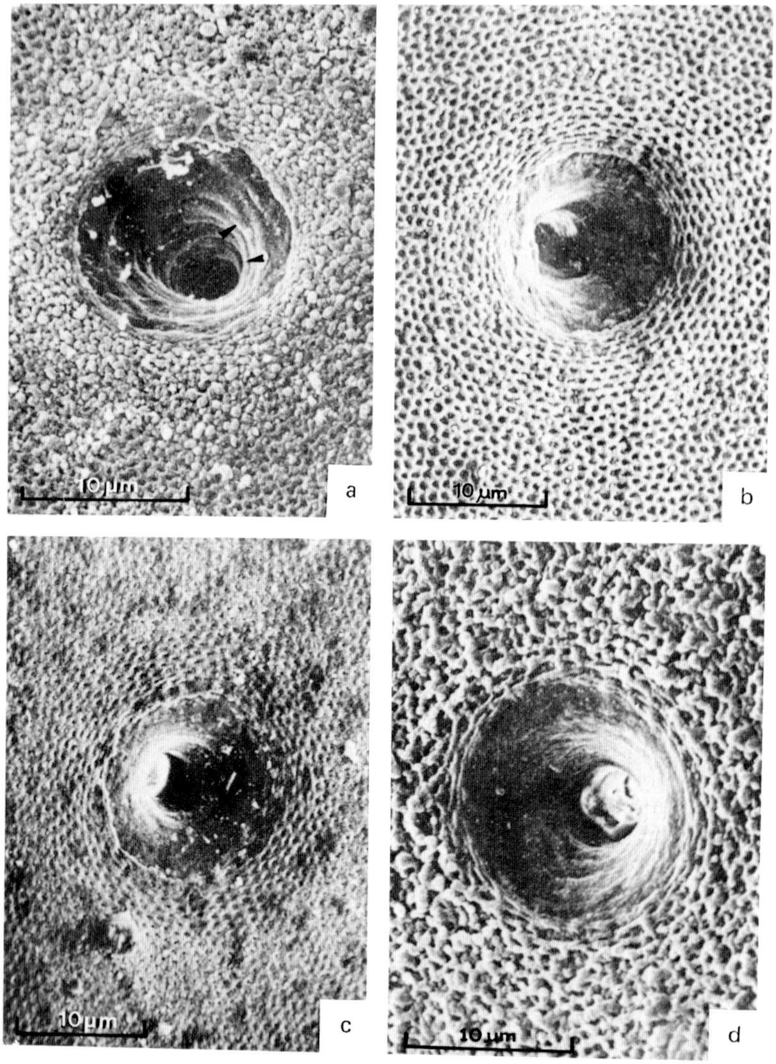

Fig. 62. Micropyles. *a* Micropyle of *Coregonus pidschian.* Look at the reinforcement of the sides of the micropyle canal (arrowheads). *b* Micropyle and surface pattern of an egg of *Coregonus nasus. c* Micropyle and surface pattern of *Goregonus oxyrhynchus. d* Micropyle of *Coregonus lavaretus.* The surface of the zona radiata is covered with minute knobs [from *Riehl,* 1980b].

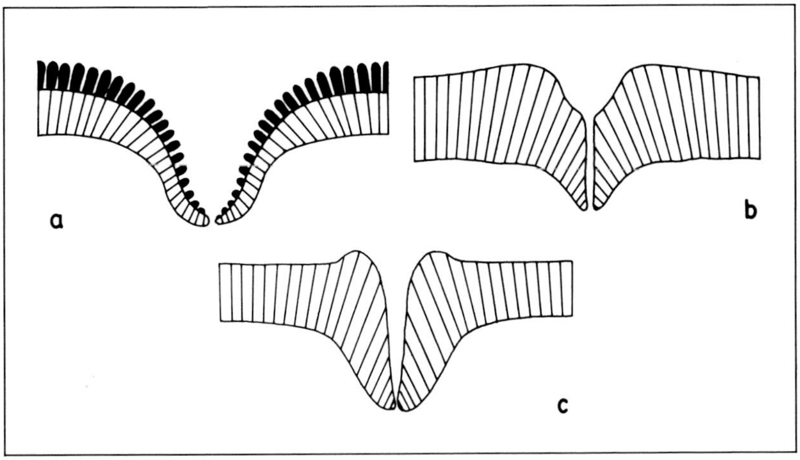

Fig. 63. Various types of micropyles: *a* type I; *b* type II; *c* type III [from *Riehl and Schulte,* 1977b].

the egg of *Brachydanio rerio* is slightly larger than the size of its sperm head, the block to polyspermy is considered to be mechanical and guaranteed by the morphological design of the micropyle [*Hart and Donovan,* 1983]. Thus, micropyles of teleost eggs facilitate fertilization [*Ginsburg,* 1968]. However, in *Esox lucius* the eggs are fertilized even when the micropyle is blocked [*Riehl and Götting,* 1975]. *Riehl and Schulte* [1977b] observed that pipe eggs can be fertilized after their micropyles are clogged, suggesting that egg envelope may be penetrated by the sperm at other sites. *Riehl* [1977e] has discussed the function of teleost micropyle in detail.

E. Theca and Surface Epithelium

1. Surface Epithelium

Surface epithelium (or serosa) surrounding the thecal layer forms the outermost layer of the follicles in the teleost ovary (fig. 42b). The serosal cells are thin, flattened, and connected with one another by tight junctions and desmosomes [*Nagahama* et al., 1978]. Their lateral plasma membranes are greatly convoluted. The cytoplasm contains both agranular and granular forms of ER; the latter are relatively more developed. The serosal cells also show small mitochondria, microfilaments, and free ribosomes. Micropinocytotic vesicles are generally present at both apical and basal surfaces. *Toshimori and Oura* [1979] reported an usual architecture of occluding junctions between surface epithelial cells in teleost *(Plecoglossus alivelis)* ovarian follicles. On the p-face, double strands of intramembranous particles are present. A narrow furrow-like gap is present between two rows of particles in these strands. On the E-face, two groove types are present. The double strands on the P-face are evidently registered with the grooves (type I or type II) on the complementary E-face and a row of particles on the E-face is registered with a furrow-like region between two rows in the double strands on the P-face.

2. Thecal Layer and Steroidogenesis

Theca constitutes a thick layer that surrounds the follicular epithelium outside the basal lamina (fig. 42b, 49, 59). The basal lamina separating the thecal layer from the follicular epithelium stains for carbohydrates and proteins [*Guraya,* 1978] and contains collagenous fibres [*Anderson,* 1967]. Electron microscope studies have revealed that it consists of a series of membranes and electron-dense material [*Hurley and Fisher,* 1966; *Anderson,* 1967; *Azevedo,* 1974; *Götting,* 1967, 1970, 1974, 1976; *Flügel,* 1967a; *Erhardt and Götting,* 1970].

Thecal cell layer is composed of collagenous fibres, capillary loops and fibroblast-like (or stromal) cells (fig. 49, 59) [*Hurley and Fisher,* 1966]. Electron microscope studies have demonstrated the presence of some enlarged (or hypertrophied) cells in the thecal layer of teleost follicles, which are generally called as the thecal gland cells or special thecal cells (fig. 48) [*Yamamoto and Onozato,* 1968; *Nicholls and Maple,* 1972; *Nagahama* et al., 1976, 1978, 1982; *Kagawa* et al., 1981; *Guraya and Kaur,* 1982; *Nagahama,* 1983; *Fostier* et al., 1983].

The vascularized theca interna in the dogfish *Scoliodon* follicle consists of large spherical or polygonal cells with abundant cytoplasm and a vesicular nucleus having a large nucleolus rich in RNA [*Guraya,* 1978]. The cytoplasm shows RNA and some diffuse lipoproteins and granular mitochondria having the usual lipoprotein composition [*Guraya,* 1978]; thecal tissue also shows lipid droplets composed of phospholipids, which are sparsely distributed. Outer to the theca interna lies the broad theca externa, consisting of fibroblast-like cells having very little cytoplasmic differentiation. The fibroblast-like cells in the thecal layer of fish follicle do not show much development of cytoplasm which contains some elements of granular endoplasmic reticulum, free ribosomes, a few mitochondria of small size, and a very small Golgi zone [*Hurley and Fisher,* 1966; *Yamamoto and Onozato,* 1968; *Azevedo,* 1974]. The stromal cells of the thecal layer are associated with each other by maculae adherens [*Anderson,* 1976; Azevedo, 1974].

Histochemical studies have demonstrated the presence of various enzymes in the thecal layer which are related either to permeability or to steroid biosynthesis [*Guraya,* 1976, 1978, 1979a; *Nagahama* et al., 1982; *Colombo* et al., 1982; *Nagahama,* 1983; *Fostier* et al., 1983]. A strong alkaline phosphatase activity is seen in the thecal layer of the fish follicle. This is believed to be involved in the transport of substances across its various membranes [*Varma and Guraya,* 1968; *Varma,* 1970]. In contrast to the observations of *Lambert* [1966, 1970a] on *Poecilia reticulata* ovaries, which have demonstrated enzyme activity related to steroidogenesis in the follicle cells, *Bara* [1965], while investigating the distribution of Δ5-3β-hydroxysteroid dehydrogenase (3β-HSDH) in the ovary of the mackerel *Scomber scomber* at different stages of the reproductive cycle, has observed that 3β-HSDH activity is greatest in some of the thecal cells and not in the follicular epithelium. Activity is strongest at the start of vitellogenesis and is decreased in intensity as the follicle matures. The distribution of glucose-6-phosphate dehydrogenase (G-6-PDH) follows the distribution of 3β-HSDH, except that it also occurs at a low level in the follicle cells [*Bara,* 1965]. In the zebra fish *(Brachydanio rerio),* 3β-HSDH activity is also seen in the thecal layer, where it is present in enlarged cells [*Yamamoto and Onozato,* 1968; *Van Ree* et al., 1977] as also reported for the goldfish [*Nagahama* et al., 1976], and carp, *Cyprinus carpio* [*Colombo* et al., 1982]. *Yaron* [1971] observed 3β-HSDH activity in both follicle and thecal cells of *Acanthobrama terrae-sancta* and *Tilapia nilotica. Saidapur and Nadkarni* [1976] also demonstrated 3β-, 17β- and 11β-HSDH and G-6-PDH activities in the thecal cells of the ovary in *Mystus cavasius* as also demonstrated in stromal cells of *Channa gachua* [*Shanbag and Nadkarni,* 1981]. The presence of 17β-HSDH has also been reported in the thecal cells of the ovaries of *Trachurus mediterraneus* [*Bara,* 1974]. The thecal and interstitial cells of the adult trout ovary show maximum 3β-HSDH activity at the time of meiotic maturation and ovulation [*van den Hurk and Peute,* 1979], leading investigators to believe that these cells may secrete progestins and/or corticosteroids, hormones involved in the regulation of these processes. However, *Kagawa* et al. [1981] did not detect 3β-HSDH activity in the special thecal cells at any stage in the preovulatory follicles of *Salvelinus leucomaenis.* These results have suggested that the special thecal cells at this stage of preovulatory growth of the follicle have low activity in synthesizing steroid hormone.

Various electron microscope studies have clearly demonstrated the presence of isolated special thecal cells lying close to capillaries in the thecal layer of ovarian follicles in different teleosts such as *Brachydanio rerio* [*Yamamoto and Onozato,* 1968], *Cichlasoma nigrofasciata* and *Haplochromis multicolor* [*Nicholls and Maple,* 1972], *Carassius auratus* [*Nagahama* et al., 1976], *Oncorhynchus kisutch* [*Nagahama* et al., 1978] and *S. gairdneri* [*Van den Hurk and Peute,* 1979], *Salvelinus leucomaenis* [*Kagawa* et al., 1981]; these cells may also be grouped in clusters in some teleosts [*Bara,* 1965, 1974; *Livni,* 1971; *Nagahama* et al., 1978] and could be of the same stromal origin as the interstitial cells [*Yamamoto and Onozato,* 1968; *Lambert and Van Oordt,* 1974; *Saidapur and Nadkarni,* 1976; *Guraya,* 1976a, 1979b; *van den Hurk and Peute,* 1979]. *Yamamoto and Onozato* [1968] have observed that special thecal cells in the zebra fish *Brachydanio rerio* originate from ovarian interstitial cells at an early stage and later become located in the thecal layer [*Guraya and Kaur,* 1982]. The hypertrophied cells, which are larger than the other thecal cells (fig. 48), show the ultrastructural features of well-established steroid gland cells [*Christensen and Gillim,* 1969; *Guraya,* 1971, 1974; *Christensen,* 1975; *Neaves,* 1975], as described for various teleost species [*Yamamoto and Onozato,* 1968; *Nicholls and Maple,* 1972; *Nagahama* et al., 1976, 1978; *van den Hurk and Peute,* 1979; *Iwamatsu and Ohta,* 1981]. Special thecal cells show a central nucleus with a conspicuous nucleolus. Agranular endoplasmic reticulum is conspicuously developed throughout the cytoplasm; granular endoplasmic reticulum is relatively less developed but often appears as several stacked cisternae (fig. 64). The moderately developed Golgi complex is generally placed close to the nucleus. Round, oval or rod-shaped mitochondria having tubular or vesicular cristae are present (fig. 56b, 64). Membrane-bound dense bodies of variable size are frequently seen near the Golgi apparatus.

The cytological and histochemical characteristics of the special thecal cells are undoubtedly indicative of steroidogenesis as also discussed in detail in previous reviews [*Guraya,* 1976a, 1978, 1979a; *Nagahama* et al., 1982; *Nagahama,* 1983; *Fostier* et al., 1983]. From the discussion of both histochemical and ultrastructural data, it can be concluded that the special thecal cells of follicular wall form the major cellular site for steroidogenesis in the teleost ovary [see also *Guraya,* 1976a, 1978, 1979a; *Hoar and Nagahama,* 1978; *Nagahama* et al., 1982; *Nagahama,* 1983, 1984; *Fostier* et al., 1983]. At present these cells are believed to secrete either 17β-oestradiol or

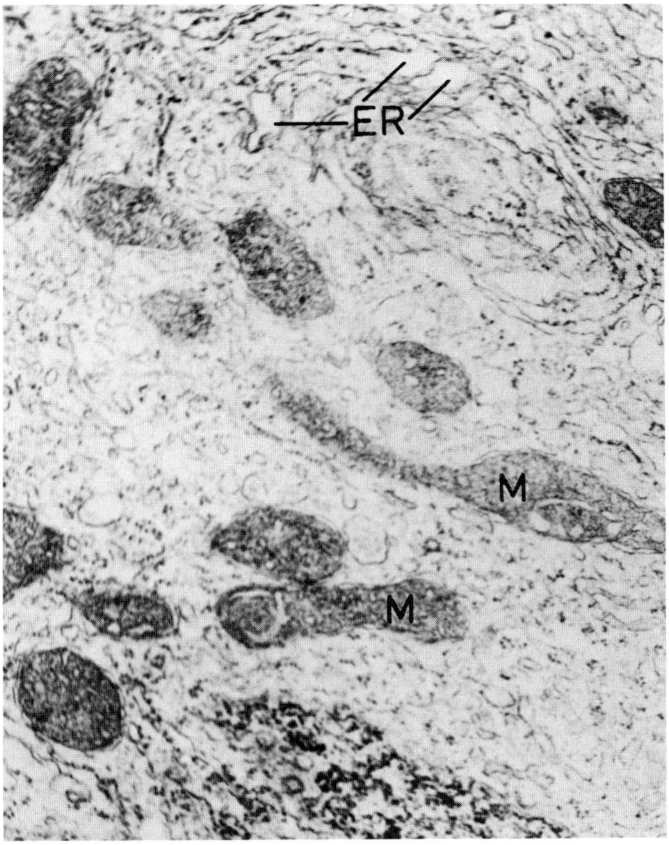

Fig. 64. Electron micrograph of a portion of a special cell of a mature coho. Note large mitochondria (M) with tubular cristae and extensive agranular endoplasmic reticulum (ER). × 20,000 [from *Nagahama* et al., 1978].

progesterone [*Nagahama* et al., 1978; *Kagawa* et al., 1981; *Nagahama*, 1983]. But the nature of steroid hormones secreted by the special thecal cells in vivo needs to be determined more precisely. *Kagawa* et al. [1981] have discussed the significance of preovulatory rise in plasma progesterone levels in relation to the induction of oocyte maturation (see also Chapter VI). But more precisely timed sampling before or during final maturation will be required to determine the cellular sources of progesterone production in the follicle wall. It will be interesting to mention here that 17α,20β-dihydroxy-4-pregnen-3-one is produced in vitro by ovarian tissue of amago salmon *(Oncorhynchus rhodurus)* and it is believed to be one of the most potent inducers of maturation [*Suzuki* et al., 1981] (see Chapter VI). The other workers have also observed that it is synthesized by the follicle in response to gonadotrophin [*Fostier* et al., 1981; *Nagahama* et al., 1983a; *Young* et al., 1983b; *Nagahama,* 1983, 1984; *Ueda* et al., 1984; *Yamauchi* et al., 1984; *Hirose* et al., 1985] and its elevated levels are present in the plasma of females undergoing final oocyte maturation [*Nagahama,* 1983; *Fostier* et al., 1983; *Young* et al., 1983b; *Ueda* et al., 1984; *Yamauchi* et al., 1984; *Hirose* et al., 1985]. The in vitro experiments have shown that under the influence of gonadotrophin, the thecal layer produces 17β-hydroxyprogesterone which is transferred to the follicular epithelium and converted to the 17α,20β-dihydroxy-4-pregnen-3-one (the maturation-inducing steroid) [*Nagahama,* 1983, 1984]; 20β-hydroxysteroid dehydrogenase, the key enzyme involved in this conversion, has been demonstrated in the granulosa layer.

Further correlative, histochemical, ultrastructural, autoradiographic and biochemical techniques, including in vitro experiments, must be used to determine more precisely the cellular sites of steroidogenesis during the maturation of the follicular wall in the fish ovary, which may be variable depending on the species and its mode of reproduction. Recent in vitro studies have provided direct evidence for the presence of 3β-HSDH in amago salmon *(Oncorhynchus rhodurus)* follicles and indicate that this enzyme is necessary for stimulation of 17β-oestradiol by mammalian and fish gonadotrophins [*Young* et al., 1982a; *Kagawa* et al., 1982a, b; *Fostier* et al., 1983]. These results indicate that fish gonadotrophin (SG-G 100) directly controls 17β-oestradiol production in amago salmon follicles although the cellular site of its synthesis and the mechanism of gonadotrophin stimulation of steroidogenesis remains unclear [see also *Fostier* et al., 1983]. However, studies on the amago salmon using isolated thecal and follicle cells have suggested that both cell types are necessary for 17β-oestradiol production in vitro in response to salmon gonadotrophin [*Nagahama* et al., 1982; *Kagawa* et al., 1982b, 1983; *Nagahama,* 1983, 1984]. In response to hCG, *Kagawa* et al. [1984] have studied in vitro 17β-oestradiol and testosterone production by ovarian follicles at different stages of their development and maturation in the goldfish, *Carassius auratus.* The decrease in 17β-oestradiol production in tertiary yolk stage follicles has been attributed partly to a decrease of aromatase activity at this stage. The

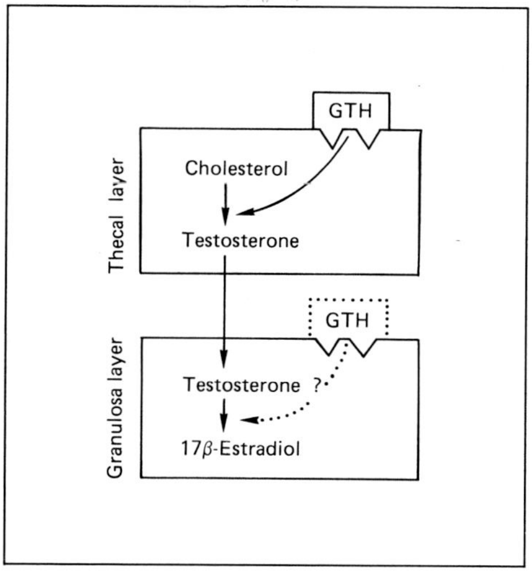

Fig. 65. Two-cell type model for the synthesis of follicular 17β-oestradiol in the amago salmon *(Oncorhynchus rhodurus)* (see the text for details) [Redrawn from *Nagahama,* 1983].

role of the vascularized thecal layer during the major part of vitellogenesis is postulated to be the production of oestrogen precursors (androgens, especially testosterone), which are converted to 17β-oestradiol in the follicular epithelium (follicle cells) by its aromatase system (fig. 65) [*Young* et al., 1982b]. Thecal layers, lacking an aromatizing enzyme system are unable to produce 17β-oestradiol; high levels of plasma testosterone in vitellogenic females are believed to serve as precursor for oestrogen synthesis [*Kagawa* et al., 1982b, 1983, 1984]. These results have suggested that both layers are needed for gonadotrophin-stimulated 17β-oestradiol production. Ultrastructural, histochemical and in vitro biochemical experiment studies have suggested that the thecal layer is responsible for androgen production involving numerous biosynthetic steps, and the granulosa layer is the site of aromatization of androgens, involving only a few biosynthetic steps. In other words, the thecal cells are ultrastructurally more steroidogenic than the granulosa cells which show fewer signs of steroidogenesis as already discussed. A similar two-cell type model for the production of follicular oestrogens in vitro has also been proposed for the rat in which granulosa cells also show fewer signs of steroidogenesis [*Guraya,* 1985]. In the rat granulosa cells, FSH plays a significant role in the aromatization of androgens to form oestrogen (or stimulates granulosa cell aromatase activity). But studies are needed to determine the involvement of gonadotrophin(s) in the conversion of androgens to 17β-oestradiol by the follicle (or granulosa) cells of the fish ovary. In other words, the mechanism of the induction or activation of the granulosa cell aromatase system is not known. Nevertheless, their aromatase activity increases during vitellogenesis and thereafter rapidly decreases in the postvitellogenic period [*Young* et al., 1983a]. The mechanism by which fish gonadotrophin stimulates testosterone production by the thecal layer also needs to be determined. Further studies are required to establish whether this pattern also occurs in vivo.

VI. Ovum Maturation

After the oocyte is fully grown during vitellogenesis, it becomes ready for the next phase of oogenesis, i.e., the resumption of reduction division (for final meiotic maturation) which is accompanied by several maturational processes in the nucleus and cytoplasm (or nuclear and cytoplasmic changes) of oocyte. These changes occur prior to ovulation and under the influence of hormonal stimuli. In this chapter, these processes or changes are collectively called as the oocyte final maturation that forms a prerequisite for successful fertilization, consists of the breakdown of the germinal vesicle (GVBD), chromosome condensation and extrusion of the first polar body. The transition from vitellogenesis to final maturation is not clearly demarcated for many fish species because oocyte growth and maturation constitute a continuous developmental process. If the oocyte final maturation is delineated at the time when oocytes respond to hormonal stimuli in vitro for undergoing GVBD, then the onset of this period varies greatly for different species and relates to the position of the germinal vesicle within the oocyte and size of the oocyte [*Goetz and Theofan,* 1979; *Wallace and Selman,* 1980]. The morphological (including ultrastructural), biochemical and endocrinological aspects of oocyte maturation in vivo and in vitro will be described briefly here.

A. Hormonal Regulation

Actually a series of complex interactions between pituitary gonadotrophin(s) and various steroid hormones secreted by the follicle wall regulate several aspects of ovarian development in fish. Teleost gonadotrophin (glycoprotein-rich gonadotrophin) is known to stimulate steroidogenesis in the follicle, oocyte maturation and ovulation. In most cases, the gonadotrophin action on various processes of follicle growth, maturation and ovulation is not direct, but through the production of follicular steroid hormones which in turn mediate vitellogenin synthesis in the liver (see Chapter IV), oocyte maturation and ovulation. Recent studies have indicated that these aspects of ovarian development are regulated by changing level and nature of steroid hormones. Hormonal regulation of oocyte maturation in numerous teleosts has been studied extensively both in vivo and in vitro experiments [see references in *Fostier and Jalabert,* 1982; *Goetz,* 1983; *Donaldson and Hunter,* 1983; *Nagahama* et al., 1983; *Richter* et al., 1982]. The detailed effects of gonadotrophins and a variety of steroids on oocyte maturation are discussed in these excellent reviews and papers and thus will be omitted here. An attempt is made here to describe briefly the various concepts of hormonal regulation of ovum maturation, which have emerged as a result of extensive in vitro and in vivo studies on different fish species. In this regard, there exist very divergent views which can be attributed partly to species variations and partly to experimental approaches (including in vitro conditions) used in different studies. Various in vitro studies have, however, shown that oestrogens are generally ineffective in inducing oocyte final maturation; these steroids may still have an indirect role in the timing of oocyte final maturation by affecting the release of gonadotrophins from the pituitary by influencing the gonadotrophin stimulation of ovarian steroidogenesis [*Fostier* et al., 1983; *Goetz,* 1983; *Nagahama,* 1984]. Androgens constitute the second least effective steroid category, followed by the 11-oxygenated corticosteroids which, however, may be quite effective in some teleosts [*Sundararaj and Goswami,* 1977; *Goetz,* 1983]. The most effective steroids are progestogens, especially 17α,20β-dihydroxy-4-pregnen-2-one (17α,20β-DHP) [*Nagahama* et al., 1983b]; 17α,20β-DHP has been identified in the plasma of several teleost species at the time of ovum maturation [*Nagahama* et al., 1983b; *Young* et al., 1983b; *Yamauchi* et al., 1984; *Ueda* et al., 1984; *Hirose* et al., 1985]. The combinations of certain steroids are more effective than the individual steroids alone. Even in those teleost species in which corticosteroids are relatively ineffective, they may play a role in oocyte final maturation, as corticosteroids are described to enhance the action of gonadotrophin and 17α,20β-DHP on GVBD. All these steroids have been shown to be produced by fish ovaries in in vitro experiments [*Fostier* et al., 1983; *Goetz,* 1983; *Young* et al., 1983b]; however, the production of corticosteroids by the interrenal has to be kept in mind. The molecular mechanisms of action of all these steroids are still to be explained. The possibility of further metabolism in the follicle of steroids, which are effective inducers of GVBD, in vitro must also be kept in mind.

As a result of in vitro studies, a great deal is known about the hormonal regulation of final oocyte maturation in fish. Much of the knowledge obtained from these in vitro studies is now being applied to in vivo situation for having an insight into hormonal control of reproduction in fish [*Donaldson and Hunter*, 1983]. Homologous or heterologous pituitary extracts (especially carp pituitary and salmon pituitary extracts), partially purified gonadotrophin preparations from preserved pituitary gland of fish (or piscine gonadotrophin), and mammalian gonadotrophins (FSH, LH and hCG) have been used in maturation experiments on different teleosts [*Jalabert,* 1982; *Goetz,* 1983; *Donaldson and Hunter,* 1983]. All these experimental approaches are generally referred to as the first-generation techniques. The second-generation techniques involve the regulation of the production and/or release of endogenous gonadotrophin, i.e. gonadotrophin-releasing hormones (Gn-RH) or luteinizing-releasing hormone (LH-RH) and antioestrogens or the provision of ovarian hormones normally produced directly or indirectly by gonadotrophin stimulation such as the progestogens and prostaglandins. The second-generation maturation inducers are considered to be of great practical significance in the manipulation of breeding time and induced breeding [*Jalabert,* 1982; *Donaldson and Hunter,* 1983]. However, the use of various maturation inducers have always suffered from a lack of standardization between workers. The results of numerous studies on the hormonal regulation of oocyte maturation in vivo have indicated that glycoprotein gonadotrophin (or maturational gonadotrophin) secreted by the pituitary during the periovulatory period is the predominant and essential regulating factor for the oocyte final maturation as evidenced by the fluctuations in its levels in the serum during this period [*Fostier* et al., 1983; *Goetz,* 1983]; it has a similar effect in the in vitro system [*Jalabert and Fostier,* 1984a, b]. The regulation of its secretion from the pituitary is under the influence of Gn-RH and the pituitary shows the highest sensitivity during the spawning period [*van der Kraak* et al., 1983]. The potent LH-RH analogues can also induce ovulation in coho salmon either alone or with a primary injection of partially purified salmon gonadotrophin [*Donaldson* et al., 1982]. *Donaldson and Hunter* [1983] have discussed the roles of Gn-RH and LH-RH in the induction of ovum maturation and ovulation in different teleost species.

The pituitary gonadotrophin(s) in teleosts is considered to be one of the major endocrine factors inducing oocyte maturation. Its mechanism of action still needs to be determined. But the results of various studies have suggested that in response to the effect of glycoprotein gonadotrophin, the follicle wall produces maturational steroid hormones especially progestogens (e.g. 17α,20β-dihydroxy-4-pregnen-3-one) (17α,20β-DHP) which are secreted in higher levels during the oocyte final maturation [*Nagahama,* 1984; *Ueda* et al., 1984; *Yamauchi* et al., 1984; *Hirose* et al., 1985] and are now considered to act on the oocyte to promote its maturation in most teleosts [*Jalabert,* 1982; *Donaldson and Hunter,* 1983; *Goetz,* 1983; *Nagahama,* 1983, 1984; *Nagahama* et al., 1983b; *Jalabert and Fostier,* 1984a, b]. Both thecal and follicular epithelium layers in amago salmon are involved in the gonadotrophin-induced follicular production of 17α,20β-DHP (fig. 66) (see Chapter V; *Nagahama* [1984]). All these studies have produced evidence that 17α,20β-DHP is the most effective steroid-mediating gonadotrophin-induced maturation in a number of teleosts and thus acts as the natural follicular mediator of gonadotrophin-induced final oocyte maturation [*Nagahama,* 1984]. This suggestion is further supported by the fact that in amago salmon cyanoketone (inhibitor of 3β-HSDH) can inhibit salmon gonadotrophin (SG-G 100) and pregnenolone-stimulated oocyte final maturation in vitro but does not block maturation induced by Δ^4-steroids such as progesterone or 17α,20β-DHP [*Young* et al., 1982a, c]. The 'follicle sensitivity' to gonadotrophins depends partly on its ability to synthesize the steroids in response to gonadotrophin. Since progestogens are directly involved in the induction of final oocyte maturation in a number of teleosts, the lack of evidence for progesterone synthesis in Japanese eel *(Anguilla japonica)* is difficult to explain particularly as 17α-hydroxyprogesterone is effective for the induction of final maturation in eel oocytes in vitro [*Yamauchi and Yamamoto,* 1982]. These data have indicated that the follicle of eel is in some way deficient and cannot synthesize maturational steroids in spite of the fact that special thecal cells possessing the ultrastructural characteristics of steroid-secreting cells are present in the follicle wall. Probably, there does not occur shift in steroidogenesis, i.e. from 17β-oestradiol to progestogens.

In Indian cat fish, *Heteropneustes fossilis,* the corticosteroids produced by the interrenal in response to the gonadotrophin are involved in oocyte maturation, as the follicle sensitivity to gonadotrophins alone in vitro is extremely low [*Sundararaj and Goswami,* 1977]. The various data have indicated that facilitation of a maturational steroid(s) originates either from the ovarian or extraovarian source. The follicle 'sensitivity' to steroids depends on the follicle size, environmental factors (including temperature, photoperiod, etc.), stage of germinal vesicle migration from the centre to the periphery of oocyte, time of treatment of the fish in daily spawning cycle

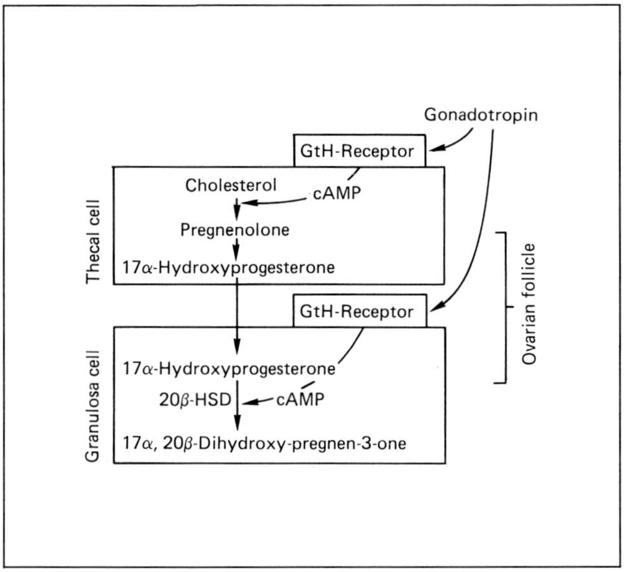

Fig. 66. Two-cell type model for the production of 17α,20β-dihydroxy-4-pregnen-3-one in the ovarian follicle of salmonids [redrawn from *Nagahama,* 1984].

[*Fostier and Jalabert,* 1982; *Goetz,* 1983]. But the molecular basis of 'follicle sensitivity' is not known. *Jalabert and Fostier* [1984a] have found that the peripheral migration of germinal vesicle (GV) in the rainbow trout coincides with an increase in oocyte sensitivity to 17α,20β-DHP. A significant correlation is observed between plasma 17β-oestradiol level and the follicular sensitivity to s-GtH (salmon maturational gonadotrophin) but not between plasma 17β-oestradiol, and the oocyte sensitivity to 17α,20β-DHP. From these results, the authors have concluded that the peripheral migration of GV is a morphological event that coincides with an increase in oocyte sensitivity and that 17β-oestradiol is the physiological regulator of follicular sensitivity to GtH. The synergistic effects of testosterone, oestrogen and cortisol on oocyte maturation need to be resolved at the molecular level. However, in vitro studies have indicated that steroids act synergistically with gonadotrophin upon oocyte maturation [*Fostier and Jalabert,* 1982; *Goetz,* 1983]. *Jalabert and Fostier* [1984b] have observed that 17β-oestradiol inhibits significantly 17α,20β-DHP output by the follicles of the rainbow trout tested at all doses of s-GTH whereas testosterone and cortisol stimulate significantly its output by follicles only at doses of GtH below 31.25 ng/ml. These observations have indicated that 17β-oestradiol, testosterone and cortisol are able to modulate the secretion of 17α,20β-DHP by trout ovarian follicles in response to GtH. The elevated plasma testosterone levels during the period of oocyte maturation in salmon are believed to be partly associated with the induction of oocyte maturation [*Kagawa* et al., 1983].

B. Morphological Characteristics

The oocyte maturation can be characterized morphologically by a clarification of the yolk as a result of the separate fusion of the proteid and lipid yolk elements, described in Chapter IV, and finally by GVBD [*Goetz,* 1983]. The fully grown oocyte of fish ovary shows a large nucleus or germinal vesicle (GV) in meiotic prophase having lampbrush chromosomes which are involved in the synthesis of ribonucleoproteins (see Chapter II). The GV of fully grown oocyte is obscured generally by the opaque ooplasm full of yolk. After treatment with certain fixing fluids, the yolk transparency is increased and the GV can be seen as a golden brown structure under transmitted light [*Goetz and Bergman,* 1978; *Goetz,* 1983]. The oocytes of goldfish take a slightly flattened shape during the later stage of vitellogenesis [*Nagahama,* 1983]. The animal pole containing the GV, on one of the flattened surfaces, is placed around the micropyle that forms a depression in the follicular epithelium and zona pellucida (see Chapter V). This area can be seen easily with the help of a dissecting microscope as it takes the form of a number of small furrows in the follicle radiating out from the micropyle [*Nagahama,* 1983]. The animal pole of the Indian catfish *(Heteropneustes fossilis)* forms a conspicuous structure with an elevated cap [*Goswami and Sundararaj,* 1971]. The gonadotrophin-induced in vitro oocyte maturation in salmonids is relatively more helpful to follow the morphological alterations of nuclear and cytoplasmic maturational changes [*Fostier and Jalabert,* 1982; *Nagahama,* 1983; *Goetz,* 1983]. With ovum maturation, the GV shifts from a central or slightly eccentric position to the animal pole where the micropyle is placed; however, the time and rate of its shifting to the peripheral position varies between species [*Goetz,* 1983]. The biophysical aspects of shifting of GV to the peripheral coplasm are not known.

The GVBD generally marks the resumption of meiosis in mammals [*Guraya,* 1985]; relatively very few studies on fish have been made to follow the sequence of chromosomal changes during resumption of meiosis. The transition from prophase I to prophase II in the Japanese medaka *(Oryzias latipes)* needs 1–4 h and occurs following GVBD both in vivo and in vitro [*Iwamatsu,* 1965; *Iwamatsu* et al., 1976]. By 5 h after GVBD in vitro, the chromosomes develop the configuration of metaphase II which, however, takes 6–7 h following GVBD in vitro in the case of pond loach *(Misgurnus angillicaudatus)* [*Iwamatsu and Katoh,* 1978]. After the completion of final maturation in vitro, the chromosomes in carp *(Cyprinus carpio)* oocytes are seen in metaphase II [*Sakun and Gureeva-Prebrazhenskaya,* 1975]. The chromosomes in various phases of the first meiotic division are also studied in some other fish species [*Yamamoto,* 1956b; *Sakun,* 1966; *Van Ree* et al., 1977].

After the migration of the GV to the periphery, its nuclear envelope breaks down (GVBD) leading to the intermixing of nuclear contents with the surrounding ooplasm. Meanwhile, there also occur cytoplasmic changes such as the coalescence of lipid yolk droplets and proteid yolk globules, a further rapid increase in the size of the oocyte caused by hydration, and an overall increase in oocyte translucency [*Wallace and Selman,* 1981; *Goetz,* 1983]. Several species variations in regard to these cytoplasmic changes are noticed. In some fish species, complete fusion of yolk globules is not seen during oocyte final maturation. Actually, degree of coalescence of yolk globules varies greatly among species belonging to different orders of fish and is considered to be related to their quantity and distribution in the ooplasm [*Goetz,* 1983]. But the degree of lipid droplet coalescence appears to show a phylogenetic pattern as discussed by *Goetz* [1983]. The molecular basis of the various forces involved in the coalescence of yolk bodies should be determined in future studies. Actually very little is known about the biochemical changes during ovum maturation [*Craik and Harvey,* 1984].

The initial ultrastructural changes that occur in the cytoplasm (3 h after treatment with corticosterone in vitro) of medaka and the loach, *Misgurnus anguillicandatus,* include disappearance of the electron-dense layer and annulate lamellae and alterations in the shape of mitochondria from long to spherical ones [*Iwamatsu* et al., 1976; *Iwamatsu and Ohta,* 1977]. In the later stages of maturation, the ooplasm shows enlarged yolk platelets and tubular endoplasmic reticulum throughout the ooplasm; cortical alveoli are placed beneath the zona pellucida. These ultrastructural changes in ooplasmic components suggest that there occur some molecular and biophysical alterations during oocyte maturation, which should be determined precisely. Future investigations must also be undertaken to determine the molecular changes in the surface (or plasma membrane) of the egg.

C. Biochemical (or Molecular) Aspects

The mechanism(s) of action of gonadotrophin(s) and steroid hormones on the follicle wall in fish are largely unknown at the molecular level. However, there occurs a shift in steroidogenesis, i.e. from production of 17β-oestradiol during exogenous vitellogenesis to progestogens during oocyte maturation [*Suzuki* et al., 1981; *Nagahama,* 1983, 1984; *Kagawa* et al., 1983, 1984; *Ueda* et al., 1984]. The nature of regulatory factors for this shift must be determined at the molecular level in future studies by using in vivo and in vitro experiments. However, in the trout the decrease of plasma 17β-oestradiol, related to a fall in aromatase activity, is believed to be controlled by gonadotrophin [*Fostier and Jalabert,* 1982]. A decrease in aromatase activity has been observed in the ovarian follicle cells of amago salmon *(Oncorhynchus rhodorus)* associated with final oocyte maturation [*Young* et al., 1983a; *Graham* et al., 1983]. A similar reduction in aromatase activity has also been reported for the postvitellogenic follicles of the goldfish, *Carassius auratus* [*Kagawa* et al., 1984] and rainbow trout, *S. gairdneri* [*Van Bohemen and Lambert,* 1981]. The molecular mechanism involved in the inhibition of aromatase activity in follicle cells is not known. *Zohar* et al. [1982a, b] have shown that the modification in ovarian steroidogenesis can be amplified by the shape and intensity of the maturational gonadotrophin. Actually, the ovarian competence and sensitivity appear to be the result of complex interaction between numerous factors including hormones at the molecular level, about which our knowledge is still meagre [*Fostier and Jalabert,* 1982; *Goetz,* 1983; *Jalabert and Fostier,* 1984]. The in vitro production of 17α,20β-DHP by the ooculture preparations is remarkably enhanced by gonadotrophin, and interaction of thecal and granulosa layers is necessary for the production of 17α,20β-DHP in response to gonadotrophin [see *Nagahama,* 1983, 1984; *Young* et al., 1983b]. Under the influence of gonadotrophin, the thecal layer produces precursors, possibly 17α-hydroxyprogesterone, which are transferred to the follicular epithelium and converted to the maturation-inducing steroid, 17α,20β-DHP (fig. 66).

The mechanism of steroid action for induction of oocyte final maturation is not known for fish. However, the involvement of cAMP, as the second messenger, in oocyte maturation has been demonstrated for amphibians [*Goetz,* 1983] and mammals [*Guraya,* 1985]. The steroidogenic response in rainbow trout is also mediated by c-AMP and its amplitude, depending on the stage, is considered to be regulated at a step beyond cAMP production [*Fostier and Jalabert,* 1983]. cAMP probably inhibits the formation within the ooplasm of maturation-promoting factor (MPF) which promotes germinal vesicle breakdown as also suggested for mammals [*Guraya,* 1985]. The precise chronology of some events of follicle maturation are defined by using inhibitors of protein and RNA synthesis [*Jalabert,* 1976]. The use of translational and transcriptional inhibitors in some studies on fish has suggested that transcription of RNA is not necessary for steroid-induced final maturation, but translation is [*Goetz,* 1983]. The cytoplasmic stimuli or signals, which regulate translation, form the promising area for future investigations. The presence of a maturational promoting factor in sturgeon oocytes has been suggested [*Masui and Clarke,* 1979]. A cytoplasmic factor (MPF) has also been found in the mature oocytes of the goldfish [*Nagahama,* 1983]. But further studies are needed on other fish species to confirm its presence as well as to determine its chemical nature. The chemical nature and site of synthesis of factor(s) that inhibit or promote oocyte maturation in the fish must be resolved in future studies as some advances in this regard have already been made for the oocyte maturation in mammals [*Guraya,* 1985]. Further studies at the molecular levels may provide the solution to the problems of follicle wall-oocyte sensitivity to gonadotrophins and steroid hormone, which appears to be related to the stage of their differentiation and maturation. Actually, cell-to-cell interactions within the ovary are of great significance to hormonize the morphological and functional differentiation of the various cell types (thecal cells, follicle cells and oocyte) in order to ensure the development, differentiation, maturation and ovulation of mature eggs which are able to be fertilized and develop normally. Some interactions are already discussed in different chapters of this monograph. The follicle cells appear to perform numerous functions through their contact with the maturing oocyte (see Chapter V). Besides producing steroid hormones, nutrients and possibly informational molecules for the oocyte, the follicle cells may also be the source of oocyte maturation-inhibiting factor(s) as also suggested for the mammalian granulosa cells [*Guraya,* 1985]. The presence of such a factor in fish is supported by the fact that there is seen increased efficiency of maturation-inducing steroid in vitro after the removal of follicular epithelium in the medaka [*Iwamatsu,* 1980; *Goetz,* 1983]. Furthermore, oocytes of this species, obtained in vivo after the endogenous stimulation necessary for further spontaneous maturation in vitro, show normal developmental activity only when surrounded by the follicular epithelium [*Iwamatsu and Ohta,* 1981b], which appears to provide the maturing oocyte with 'special factors' dependent upon gonadotrophic stimulation. The gonadotrophin stimulates the synthesis and/or release of a maturational steroids from the follicular epithelium which has been shown to contain enzyme activities related to steroid biosynthesis (see Chapter V).

Further correlative histochemical, ultrastructural, biophysical and biochemical studies during ovum maturation will be very useful to determine the precise nature of subcellular and molecular alterations that occur in the components of follicle wall and oocyte. The changing pattern of stage-specific synthesis of proteins and RNAs described for mammals [*Guraya,* 1985] needs to be resolved during oocyte growth and maturation in the fish. This will help us to determine the chronological origin and nature of cytoplasmic factors which inhibit and promote oocyte meiotic maturation. A thorough understanding of molecular mechanisms involved in the regulation of meiotic divisions in fish oocyte will be very useful to develop better chromosome engineering techniques for the sex control and genome manipulation in early embryology of the fish [*Thorgaard,* 1983] as not much research work has been carried out previously on the inactivation of egg chromosomes to induce androgenesis. This is due to the fact that radiation used for this purpose not only damages the DNA of egg chromosomes but also possibly affects the mitochondrial DNA, mRNA and other constituents of the egg cytoplasm. Physical treatments of eggs with cold shock, pressure, etc., to induce androgenesis are believed to damage egg constituents less than radiation treatment. But the molecular basis of these differences needs to be understood in future studies. The studies of sex control and of natural sex reversal mechanism(s) in fishes are of great economic importance in the development of aquaculture and fish farming on a scientific basis [*Chan and Yeung,* 1983; *Thorgaard,* 1983; *Hunter and Donaldson,* 1983].

VII. Ovulation

The ovulation is the process by which the eggs after the completion of the first meiotic division are released into the ovarian or peritoneal cavity. The release of the mature egg through a specific area of the follicle (often referred to as the stigma) is apparently not a passive process in which the follicle wall simply disintegrates, leaving the egg free from the follicle wall. The various experimental studies have indicated that expulsion of the egg takes place by the contraction of the follicle wall pushing the oocyte out (expulsion). Actually, very divergent views exist about the mechanism of ovulation that has been more extensively studied in mammals [*Guraya,* 1985] than in fish [*Nagahama,* 1983; *Goetz,* 1983] and other vertebrates [*Espey,* 1978]. The various factors such as blood vascularity, pressure, muscular, nervous, neuromuscular, proteolytic enzymes, etc., have been suggested to be involved in mammalian ovulation [*Guraya,* 1985]. Not a single factor is sufficient to explain the mechanism of ovulation but a combination of factors appear to be involved in ovulation. A similar suggestion can also be made for the ovulation in fish.

The meiosis continues up to the second meiotic metaphase in the fish ovulated eggs which then become ready for fertilization. Actually, both the meiotic maturation and ovulation constitute the end-stages of oogenesis, which are regulated by pituitary hormones (gonadotrophin(s)) and steroid hormones secreted by the follicle wall [*Fostier and Jalabert,* 1982; *Goetz,* 1983; *Guraya,* 1985]; the respective roles of these hormones are also controversial in fish in spite of the numerous in vitro and in vivo studies made so far [*Goetz,* 1983; *Donaldson and Hunter,* 1983].

The electron microscope studies have shown that the follicle cell processes and oocyte microvilli described in Chapter V are withdrawn from the zona pellucida (or chorion) during ovulation leading to the formation of a wide space between the follicular epithelium and the zona pellucida (fig. 54, 55) [*Flügel,* 1967a, b; *Yamamoto and Yamazaki,* 1967; *Hirose,* 1972; *Pendergrass and Schroeder,* 1976; *Iwamatsu and Ohta,* 1977]. The nature of factor(s) involved in the interruption of follicle-oocyte communication prior to ovulation is not known. *Oshiro and Hibiya* [1975, 1982] have suggested the involvement of proteolytic enzymes in the disruption of follicle oocyte interrelationships. The early phase of RNA synthesis is considered to be involved in the production of such enzymes needed for follicle-oocyte disruption and/or in the formation of a hole in the follicle [*Theofan and Goetz,* 1981]. The cell types of the follicle wall involved in the synthesis of these enzymes need to be determined. In the trout, the disruption of oocyte-follicle communication occurs during oocyte maturation and requires stimulation by high levels of 17α-hydroxy-20β-dihydroprogesterone in vitro [*Jalabert,* 1978, 1982]. It is possible that the agents responsible for the rupture are distinct from those involved in follicular separation and therefore may be under separate hormonal regulation. The precise role of intraovarian and extraovarian factors (or hormones) in the synthesis of proteolytic enzymes needs to be worked out.

Prostaglandin $F_{2\alpha}$ can induce ovulation in vitro of goldfish oocytes provided these had undergone maturation in vivo after treatment with hCG [*Kagawa and Nagahama,* 1981]. In response to prostaglandin, the oocytes are released into the incubation medium through a rupture in the follicle wall at the point of attachment to the overlying epithelium. The ovulation in the trout can also be induced in vitro by prostaglandin $F_{2\alpha}$ provided its ovaries were exposed to the action of 17α-hydroxy-20β-dihydroprogesterone in vivo [*Jalabert* et al., 1978]. *Goetz* [1983] has discussed in detail the roles of prostaglandins which appear to be natural mediators of oocyte expulsion in fish as also described for mammals [*Guraya,* 1985]. Of the various prostaglandins tested, $PGF_{2\alpha}$ is usually the most effective in follicular rupture in fish. The prostaglandins are considered to be involved in ovulation possibly by stimulating follicular contraction. The indomethacin, an anti-inflammatory agent, which blocks prostaglandin synthesis from dihomo-γ-linolenic acid and arachidonic acid, inhibits ovulation in fish [*Goetz,* 1983; *Donaldson and Hunter,* 1983] and mammals [*Guraya,* 1985]. This inhibition can be overridden by injection of exogenous prostaglandin in fish [*Kapur,* 1979] and mammals [*Guraya,* 1985]. There is also some evidence for the action of prostaglandins at the hypothalamic-hypophysial site in the fish [*Singh and Singh,* 1976]. But the precise nature and site of action of prostaglandin(s) involved in ovulation needs to be determined; phylogenetic differences may exist in regard to the nature of prostaglandin(s) involved in ovulation. Very divergent views have also been expressed about the nature and amount of prostaglandin(s) needed for causing rupture of the follicle as the levels of prostaglandins in the plasma and ovarian tissue vary before and after ovulation in different fish species [*Goetz,* 1983].

The rupture of the follicle is closely accompanied by the contraction of its wall [*Fostier and Jalabert,* 1982; *Nagahama,* 1983]. This contraction may be brought about by the microfilaments which have been demonstrated with the electron microscope, in the thecal cells at the time of ovulation [*Szöllösi and Jalabert,* 1974; *Nagahama* et al., 1976; *Pendergrass and Schroeder,* 1976]. A similar suggestion has also been made for mammals [*Guraya,* 1985]. The thecal cells of the rainbow trout contain smooth muscle-like cells [*Szöllösi and Jalabert,* 1974; *Szöllösi* et al., 1978]. The sparsely distributed 5–7 nm microfilaments, seen in the thecal cells of the medaka follicle at the start of incubation with hydrocortisone, form bundles after 10 h [*Pendergrass and Schroeder,* 1976]. The addition of cytochalasin B (an inhibitor of microfilamentous contractile elements) into the incubation medium completely inhibits hydrocortisone-induced ovulation in vitro of the medaka [*Schroeder and Pendergrass,* 1976] and partially prevented prostaglandin-induced ovulation in vitro of the rainbow trout [*Jalabert and Szöllösi,* 1975]. The role of muscular contraction in ovulation is further supported by the observation that in brook trout oocytes that have undergone GVBD and follicular separation, in vivo dibutyryl cyclic 3′,5′-adenosine monophosphate and phosphodiesterase inhibitors inhibit ovulation in vitro in incubates with or without prostaglandins [*Goetz* et al., 1982]; cyclic AMP is known to inhibit contraction of smooth muscle by interfering with phosphorylation of myosin [*Adelstein and Klee,* 1980].

All these observations have suggested that microfilaments of the thecal layer may be playing an important role of contraction for facilitating the release of egg from the follicle as also suggested for mammals [*Guraya,* 1985]. The presence of contractile cells in the follicle wall might not be needed in those fish species where the oocyte maturation is accompanied by conspicuous swelling of the oocyte and/or the production of a jelly layer by the follicle cells, and where the hydration process can provide the force to expel the oocyte as proposed in plaice [*Oshira and Hibiya,* 1981]. *Yorke and McMillan* [1980] using electron microscopy have proposed that expulsion of lamprey *(Petromyzon marinus)* egg is closely accompanied by an alteration in the shape of specific follicle cells from a cuboidal to columnar configuration. The change in shape constricts the follicle at the one end and pushes the oocyte out at the opposite end of the follicle. A similar mechanism of follicular constriction by alterations in follicle cell shape may also exist in teleosts. In conclusion, it can be stated that although the precise mechanics of ovulation are still not known, there is need for the local synthesis and action of prostaglandins [*Ogata* et al., 1979], and possibly a mediation by catecholamines [*Jalabert,* 1976; *Goetz,* 1983], suggesting a role for an intraovarian nervous innervation which, however, needs to be studied in fish ovary. But nervous innervation of mammalian ovaries has been demonstrated in several species [*Guraya,* 1985]. The in vitro results obtained by *Jalabert* [1976] suggest that epinephrine stimulates ovulation in rainbow trout through interaction with α-adrenergic receptors; the mechanism following this activation needs to be determined, although epinephrine can also stimulate in vitro ovulation in yellow perch, brook trout and carp oocytes [*Epler,* 1981; *Stacey and Goetz,* 1982; *Goetz,* 1983]. However, *Stacey and Goetz* [1982] have suggested that the epinephrine effect on yellow perch ovulation may be due to increased production of prostaglandin as evidenced from its higher levels in incubates containing epinephrine than in controls. The physiological meaning of catecholamines, which theoretically may come from sympathetic nerve endings in the ovary or from chromaffin tissue in fish ovulation, is still not clear. Further studies at the subcellular and molecular levels should be continued to determine the precise natue of factors which are involved in ovulation as our knowledge in this regard is still fragmentary.

VIII. Follicular (or Oocyte) Atresia

Follicular atresia is a degenerative process by which oocytes in various stages of their development and differentiation are lost from the ovary [*Guraya,* 1973]. It has been studied extensively in cyclostomes, elasmobranchs and teleosts [*Browning,* 1973; *Saidapur,* 1978; *Guraya,* 1979b]. Follicular atresia in the fish ovary occurs commonly during the prespawning, spawning and postspawning periods [*Guraya* et al., 1975, 1977; *Guraya,* 1979b; *Saidapur,* 1978]. Mostly, atresia affects the vitellogenic and fully mature yolky eggs. However, atresia of previtellogenic follicles has also been observed [*Guraya* et al., 1975, 1977]. For the study of atresia in the previtellogenic oocytes, a very careful examination of fish ovary is needed as these are small in size. Most of the previous studies deal with the histological changes in follicular atresia during the annual reproductive cycle and under various experimental conditions such as hypophysectomy, administration of steroid hormones, treatment with biocides, etc. Various components such as the nucleus (or germinal vesicle), ooplasm, follicular epithelium and thecal layer of previtellogenic and vitellogenic oocytes described in Chapters II–V undergo conspicuous morphological and histochemical changes which are related to their degeneration and absorption. The possible factors causing follicular atresia will also be discussed.

A. Atretic Previtellogenic Follicles

1. Morphological and Histochemical Changes

In the normal situation, the atresia of previtellogenic follicles is relatively of rare occurrence in most of the teleosts investigated [*Guraya* et al., 1975, 1977; *Saidapur,* 1978; *Babu and Nair,* 1983]. Under some situations, it is so common that almost all the previtellogenic follicles and early vitellogenic follicles are affected by atresia. For example, in some specimens of *Xenentodon cancila,* most of the previtellogenic follicles in different stages of their growth and differentiation were affected by atresia (fig. 67a–d).

Interfollicular tissue of such ovaries is highly vascularized as evidenced by the presence of numerous vessels containing blood cells. Actually interfollicular spaces were almost choked with blood vessels.

The first sign of atresia in the previtellogenic follicles is the shrinkage of oocyte, which is closely accompanied by the development of clear spaces in the peripheral ooplasm (fig. 67a–d). The ooplasm may also be separated from the nucleus by a clear space produced as a result of shrinkage. Meanwhile, there also occurs the disorganization of ooplasmic and nuclear components [*Guraya* et al., 1975, 1977; *Mani and Saxena,* 1985]. The nucleoli, which are generally arranged to form a regular layer in the nuclear envelope of normal growing previtellogenic oocyte (fig. 1), become distributed irregularly in the nucleoplasm of shrunken germinal versicle with the advancement in degenerative changes (fig. 67c). The ooplasmic organelles such as yolk nucleus substance, mitochondria, Golgi bodies, lipid bodies, etc., described in Chapter III, may form masses or aggregations of bizarre appearance, which also become highly sudanophilic as evidenced by the development of more lipids (or triglycerides) due to unmasking.

With the initiation of atresia, the follicle cells hypertrophy and start invading the shrunken nucleo-ooplasmic mass (fig. 67b) [*Guraya* et al., 1975, 1977; *Mani and Saxena,* 1985], thus acting as phagocytes. The theca of large previtellogenic follicles is also hypertrophied. The gradual advancement of atresia in the previtellogenic follicles of *M. tengara* and *Cyprinus carpio* [*Guraya* et al., 1975, 1977] and *X. cancila* [*S.S. Guraya,* unpubl. observations] results in the removal of a nuclear and ooplasmic contents, and the degenerating oocytes become taut with phagocytes derived from the follicle cells. Blood cells also invade the atretic oocyte, possibly helping in

---

Fig. 67. Histological changes during atresia of previtellogenic and possibly early vitellogenic follicles in the ovary of *Xenentodon cancila. a* All the previtellogenic follicles are affected by atresia as evidenced from early, intermediate and late stages of their degenerative changes. *b* High-power view of atretic previtellogenic follicles, showing their shrinkage and disorganization as evidenced from the presence of clear spaces (arrowheads) between the nucleus (N) and ooplasm as well as between the ooplasm and follicle wall. Note the presence of nodules of stromal tissue (NS) formed during late atresia. *c* Higher power view of atretic previtellogenic follicle, showing vacuoles (V) in the peripheral ooplasm and irregularly distributed nucleoli (NU). A lot of blood vascularity (BV) is seen in between the atretic follicles. *d* Higher power view showing irregularly distributed nucleoli (NU) and clear spaces outer to

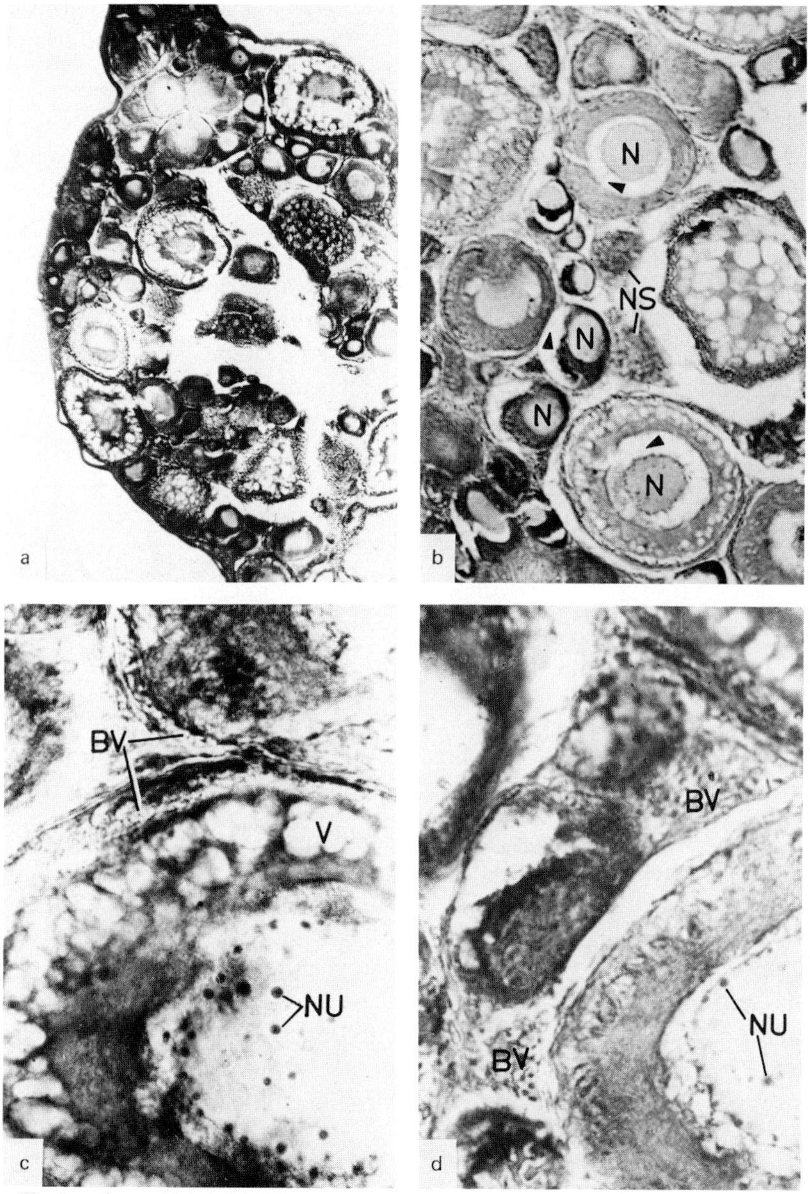

the nuclear envelope of atretic large previtellogenic follicle, which is shrunken and disorganized. Note the presence of blood vascularity (BV) in between the atretic follicles. Iron haematoxylin preparation. *a* × 37; *b* × 75; *c, d* × 300.

the resorption. The invasion of oocyte by the follicle cells is also observed in goldfish [*Beach,* 1959], *Mystus tengara* and *Amphipinous cuchis* [*Rastogi,* 1968b, 1969b]. However, *Rajalakshmi* [1966] observed that the atresia of previtellogenic follicles is hardly accompanied by the invasion of the oocyte by follicle cells. *Barr* [1968b] stated that because of lack of any distinguishing characteristic, it has proved difficult to be certain of the origin of cells which actually invade the oocyte. Thecal and follicle cells may be involved in addition to phagocytic cells derived from the ovarian stroma. After complete resorption and disappearance of the follicle contents, there is left behind in the ovary a nodule of stromal elements (fig. 67a, b), which in the case of large previtellogenic follicles show some highly sudanophilic cells of apparently thecal gland origin in histochemical preparation; such thecal gland cells in the wall of maturing follicle are described in Chapter V. Their sudanophilic lipids consist of triglycerides, cholesterol and/or its esters, and some phospholipids. These cells contain cholesterol-positive lipid droplets and constitute the interstitial cells which are distributed singly or in groups in the ovary of carp *(Cyprinus carpio)* [*Guraya and Kaur,* 1982]. During follicular atresia in the catfish, *Mystus cavasius, Saidapur and Nadkarni* [1976] also observed that the special thecal cells remain in the stroma to form the interstitial gland cells.

2. Origin and Nature of Interstitial Cells

Actually very divergent views have been expressed about the origin, distribution and function of the interstitial cells in the fish ovary [*Guraya,* 1973, 1976a, 1979b; *Saidapur,* 1978; *Guraya and Kaur,* 1982]. In the earlier studies carried out with histological techniques, the presence of interstitial cells in the fish ovary was almost denied. For the first time, *Guraya* [1972], by applying histochemical tests for lipids to frozen sections, demonstrated the origin, distribution and histochemical characteristics of interstitial cells in the dogfish *(S. sorrakowah)* ovary. Their origin is clearly related to the atresia of large previtellogenic follicles as these interstitial cells originate by the 'luteal-like' transformation of theca interna and surrounding stromal elements of atretic previtellogenic follicles as also subsequently confirmed for the teleost *Mystus cavasius* with histochemical techniques for enzymes related to steroidogenesis [*Saidapur and Nadkarni,* 1976]. A similar origin of interstitial cells has also been reported in the ovaries of different groups of nonmammalian vertebrates including some teleost fish [*Guraya,* 1973, 1976a; *Saidapur,* 1978]. The patches of interstitial cells lying in the wall of atretic previtellogenic follicles of the dogfish ovary are vascularized and contain sudanophilic lipid droplets consisting of triglycerides, cholesterol and/or its esters and some phospholipids and diffuse lipids composed of mainly lipoproteins; the latter are suggested to be derived from the abundant membranes of agranular endoplasmic reticulum of ovarian interstitial cells [*Guraya,* 1973, 1978; *Guraya and Motta,* 1980; *Guraya and Kaur,* 1982]. The interstitial cells in the ovaries of juvenile rainbow trout *(Salmo gairdneri)* are found embedded in the mesenchymal stroma at a very early stage of ovarian development (12-week-old fish) (fig. 68a) and show abundant vesicular or tubular smooth endoplasmic reticulum, round or oval mitochondria with tubular cristae, scattered ribosomes and a few membrane bound dense bodies (fig. 68b) [*Upadhyay,* 1977]. These interstitial cells of juvenile fish, which also show enzyme activities related to steroidogenesis (3β-HSDH and G-6-PDH) in the developing ovary of 3-day to 7-month-old female platyfish, *Xiphophorus maculatus* [*Schreibman* et al., 1982], can be designated as the primary interstitial cells in contrast to the secondary interstitial cells which derive from the thecal cells of atretic previtellogenic follicles in the mature fish ovary. Such a distinction has also been made for the comparable interstitial cells of mammalian ovary [*Guraya,* 1978; *Guraya and Motta,* 1980]. The interstitial cells of fish ovary contain hydroxysteroid dehydrogenases, such as 3β-, 17β- and 11β-HSDH and G-6-PDH, which are associated with steroid hormone synthesis and/or metabolism [*Guraya,* 1976a, 1979b; *Saidapur and Nadkarni,* 1976; *Saidapur,* 1978]. The physiological significance of the interstitial cells in the fish ovary remains doubtful as in the swordtail, these show cytochemical characteristics of steroid metabolism, but in the zebra fish these interstitial cells lack G-6-PDH activity. The thecal gland cells described in the wall of the maturing follicle (see Chapter V) may be representing the interstitial cells having the cytological and histochemical features of steroid-secreting cells [*Guraya and Kaur,* 1982]. These special thecal (or interstitial) cells, which show conspicuous fluctuations in their distribution, are believed to be the only cells responsible for steroid production and/or metabolism in the thecal layer of teleost follicle (see Chapter V). But the nature of steroidal secretions of fish ovarian interstitial cells is still to be determined. The secretion of androgens has been attributed to the special thecal (or interstitial) cells (see Chapter V). Further correlative electron microscope, histochemical, and biochemical studies are needed to determine the precise origin, distribution and function of interstitial cells in the fish ovary, in relation to the effects of enteroceptive and exteroceptive factors which may be regulating their histophysiology.

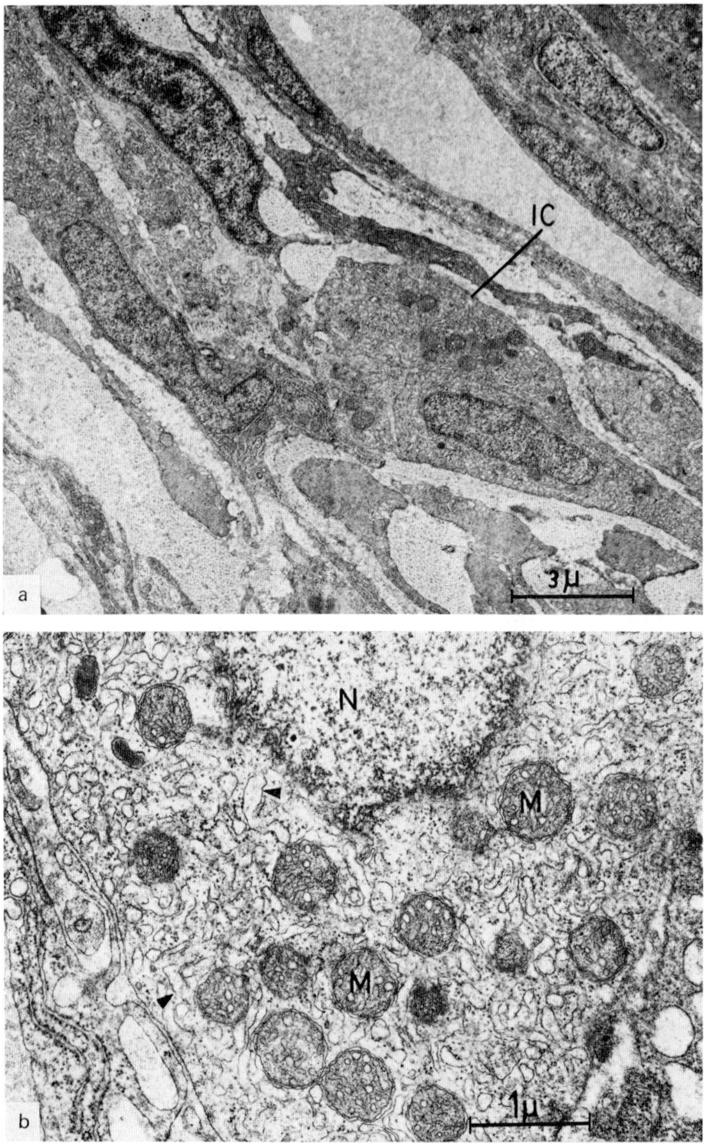

Fig. 68. Electron micrographs showing interstitial cells in the ovary of juvenile rainbow trout treated with s-GTH (0.1 μg/g body weight). *a* Interstitial cell (IC) embedded amongst the fibroblasts and collagen fibres of ovarian stroma. *b* Higher power view of portion of interstitial cell showing a vesicular nucleus (N), abundant smooth endoplasmic reticulum (arrowheads) and mitochondria (M) with tubular cristae [courtesy of Dr. *S.N. Upadhyay*].

B. Atretic Vitellogenic Follicles (or Yolky Oocytes)

1. Morphology and Histochemistry

The atresia of vitellogenic follicles especially of mature yolky eggs is of common occurrence in the fish ovary during the prespawning, spawning and postspawning periods and thus has been studied extensively in numerous species of teleosts, elasmobranchs and some species of cyclostomes [*Ball,* 1960; *Browning,* 1973; *Babu and Nair,* 1983; *Guraya* et al., 1975, 1977; *Saidapur,* 1978]. There is considerable variation in the events associated with the atresia of vitellogenic follicles, in the fish ovary. However, the main histological features are: (1) shrinkage and distortion of the entire follicle; (2) folding and dissolution of zona pellucida; (3) disorganization and liquefaction of yolk and ooplasm; (4) hypertrophy and phagocytotic character of follicle cells; (5) ingestion and digestion of yolk by the follicle cells, and (6) infiltration of theca and blood cells forming a fibrous nodule with pyknotic nuclei. A number of workers have reported the phagocytotic nature of follicle cells which play an active role in the ingestion and digestion of yolk during follicular atresia [*Guraya,* 1973; *Guraya* et al., 1975, 1977]. *Kosmath* et al. [1983a, b] using electron microscopy have reported the origin of phagocytic cells from the thecal (or stromal) cells in the atretic vitellogenic follicles of *Myxine glutinosa.* In the atretic follicle of the open involutionary type, an opening is formed in its wall, through which granulosa cells and yolk platelets are emitted. The migrating cells of the thecal layer invade follicular lumen and absorb phagocytotically residual degenerating granulosa cells. In atretic follicles of the closed involutionary type, the yolk platelets and granulosa cells remain in the follicle and phagocytotically absorbed by migrating cells of thecal origin. The secretion of yellow pigment granules during the later stages of atresia in the vitellogenic follicles has been described in different teleost species [see references in *Guraya* et al., 1975, 1977].

2. Origin and Function of Preovulatory Corpora lutea (or Corpora atretica)

Because of the histological features of an epithelial gland and accumulation of yellow pigment granules, atretic vitellogenic follicles have been generally designated as preovulatory corpora lutea or corpora atretica [*Ball,* 1960; *Lambert,* 1970b; *Browning,* 1973; *Guraya,* 1973, 1976a, 1979b; *Guraya* et al., 1975, 1977; *Babu and Nair,* 1983]. Many investigators were impressed by the glandular appearance of the teleost corpora atretica in the histological preparations and on this basis ascribed the endocrine role (steroid biosynthesis) to these ovarian structures [*Ball,* 1960; *Browning,* 1973]. However, *Hisaw and Hisaw* [1960] suggested that the preovulatory corpora lutea (corpora atretica) are probably not the endocrine gland but only the outcome of hypertrophy of follicle cells involved in the phagocytosis of yolk as also supported subsequently by *Hoar* [1965]. In order to assess endocrine function of corpora atretica, atretic vitellogenic follicles have been submitted to histochemical techniques for lipids and enzymes which are indicative of steroidogenesis in the well-established steroid gland cells [*Guraya,* 1976a, b, 1978; *Saidapur,* 1978].

Guraya and Kaur [1982], using histochemical techniques for lipids, have observed that the first change, which occurs during the atresia of ripe yolky eggs in the carp *(C. carpio),* is the development of large sudanophobic vacuoles in the region of cortical vacuoles. As the atresia of yolky eggs advances, sudanophilic fatty yolk elements consisting of triglycerides coalesce to form highly sudanophilic masses in the corpora atretica. The yolky content of such atretic vitellogenic follicles is gradually digested and removed by the follicle cells, which, at the same time, also store sudanophilic lipids consisting mainly of neutral fats (triglycerides). In the later stages of resorption, atretic vitellogenic follicles may also store cholesterol and/or its esters [*Guraya and Kaur,* 1982] as also reported for the atretic follicles of some elasmobranchs and teleosts [*Guraya,* 1976a; *Saidapur,* 1978]. These dense cholesterol-positive lipid accumulations have been attributed to degenerative changes of vitellogenic follicles [*Guraya and Kaur,* 1982]. 3β-HSDH activity could not be demonstrated in atretic follicles of different teleosts studied so far [*Guraya,* 1976a, 1979b; *Saidapur,* 1978; *Nagahama,* 1983]. However, in some teleosts, the vitellogenic follicles in early stages of atresia may show some steroid dehydrogenase activity, which may be related to hypertrophy (or incipient luteinization) of follicle cells.

The corpora atretica or preovulatory corpora lutea of lower vertebrates are usually believed to secrete steroid hormones [see references in *Browning,* 1973]. But *Guraya* [1976a] believes that these are merely the large yolky eggs in the process of their resorption. This suggestion was further supported by histochemical studies of atretic vitellogenic eggs in *C. carpio* [*Guraya and Kaur,* 1982] and other teleosts [*Nagahama,* 1983], as well as by the correlative histochemical and electron microscopic data of atretic vitellogenic follicles in the ovary of *Myxine glutinosa* [*Kosmath* et al., 1983a, c]. The various cytochemical and ultrastructural studies have also shown that the corpora atretica of teleost ovary are merely the large yolky eggs in the process of degeneration and resorption [*Guraya,* 1973, 1976a, 1979b; *Nagahama,* 1983; *Saidapur,* 1978]. However, in vitro studies are needed to substantiate this suggestion of *Guraya* who has denied the endocrine role of atretic vitellogenic follicles (or corpora atretica) in different groups of nonmammalian vertebrates including fish [*Guraya,* 1973, 1976a]. The in vitro studies on the steroidogenic capacity of atretic follicles of the goldfish have shown that incubation of its atretic follicles (to B stages according to *Khoo* [1975]) with gonadotrophin did not show a demonstrable (more than 30 pg/ml) synthesis of progesterone, testosterone, and 17β-oestradiol [*Nagahama,* 1983]. But further in vitro studies on corpora atretica of different teleosts are needed to conclude whether these ovarian structures are related only to degeneration and resorption of yolk or additionally with some other function.

In conclusion, it can be stated that atresia of vitellogenic follicles forms a common feature of fish ovary. Their follicle cells hypertrophy and become phagoytic to digest and remove ooplasmic contents; however, in some species, thecal cells may also perform similar functions during atresia. Morphological and histochemical data do not support the endocrine function of atretic vitellogenic follicles (or corpora atretica). But further in vitro incubation studies are needed to reach some definite conclusion about their endocrine function.

C. Causes and Significance of Follicular Atresia

The follicles (or oocytes) in various stages of growth in the fish ovary are lost through atresia during any time of the year. The atresia of yolky eggs forms the most conspicuous feature of fish ovary in the prespawning and postspawning periods. After breeding is over, the residual eggs are generally lost by undergoing atresia. The mechanisms, which initiate and regulate atresia of oocytes, are poorly understood at the molecular level. The various factors influencing follicular atresia have been suggested in the previous literature [*Guraya,* 1973]. These include age, stage of the reproductive cycle, hormones of extraovarian and intraovarian sources, captivity, light, temperature, crowding, nutrition, ischaemia, irradiation, etc. Recent studies have shown that chemical agents like biocides also influence follicular atresia. Some insight has been gained recently into the intraovarian and extraovarian factors which determine the ultimate fate, ovulation and atresia of follicles in the fish ovary. It is suggested that atresia is the inevitable outcome of developmental process in the follicle and that all follicles would age and become atretic unless rescued at a critical stage of development by appropriate hormonal stimuli. The interaction of both gonadotrophins and steroids is known to play an important role in the regulation of follicular development [*Nagahama,* 1983, 1984; *Fostier* et al., 1983; *Goetz,* 1983]. Several lines of evidence have shown that process of atresia in the fish ovary is either due to the lack of proper gonadotrophic stimulation or due to imperfect balance of various hormones including steroids [*Guraya,* 1973] as also suggested for mammals [*Guraya,* 1985]. The atresia of vitellogenic follicles (or yolky eggs) during the prespawning and spawning periods has been related to inhibition of gonadotropin production at the tonic level [*Saidapur,* 1978]. *Sundararaj and Goswami* [1968] observed that follicular atresia was accentuated following hypophysectomy and considered that gonadotrophin must be maintained at tonic level even when the ovary contains largely vitellogenic and yolky oocytes. Further, *Hoar* [1965] and *Barr* [1968b] have considered that the oocytes undergo atresia in case pituitary is removed after vitellogenesis has set in the oocytes, whereas only the ovulation process is inhibited when the pituitary is removed close to the onset of spawning season. Oestrogens, with oestradiol being the most potent, secreted by the follicle wall (see Chapter V), appear to have direct protective effects on follicles as evidenced by their ability to delay atresia induced by hypophysectomy [*Sundararaj and Goswami,* 1968; *Anand and Sundararaj,* 1974]. The experimental data obtained after hypophysectomy [*Chester Jones and Ball,* 1962; *Barr,* 1968b; *Saidapur and Nadkarni,* 1973] or after inhibiting the synthesis and/or release of gonadotrophins with antigonadotrophic substances, clearly support the role of gonadotrophins in follicular growth and atresia. This is probably carried out through regulation of oestrogen synthesis by the gonadotrophins as the oestrogens have direct protective effects on follicles as well as regulate vitellogenesis as discussed in Chapter IV. An antiserum to vitellogenic hormone (gonadotrophin) arrests vitellogenesis and induces follicular atresia in vitellogenic flounder [*Ng* et al., 1980a], an antiserum to maturational hormone does not affect the histochemical structure of the ovary [*Ng* et al., 1980b]. Increase in the incidence of follicular atresia in the ovaries of fishes exposed to short photoperiods is explained by the increased release of melatonin and similar substance(s) in the absence of photoperiodic stimuli which in turn inhibited the tonic production of gonadotrophin(s) [*Saxena and Anand,* 1977].

Most of the fish species investigated are seasonal breeders and bred at a time when environmental conditions such as light, temperature, water, food supply, etc., are favourable for the survival of maximum young ones [*Wootton,* 1982; *Lam,* 1983]. The initiation and course of oogenesis is controlled by environment and endocrinological factors. The vitellogenesis is the main developmental phase and in captive teleost oocytes remain at the early yolk globule stage. Their final maturation needs exogenous hormone. But without hormone stimulation, the captive grey mullet *(Mugil cephalus)* will not complete the final oocyte maturation in captivity, and the oocytes undergo atresia and degenerate [*Kuo,* 1982]. The increased atresia in the ovaries of *Heteropneustes fossilis* [*Sehgal and Sundararaj,* 1970] as well as in *Mystus tengara* [*Saxena and Anand,* 1977] has been recorded following exposure to short photoperiods (9 h/day). The role of pathogens (bacteria, fungi, helminths, etc.) in the follicular atresia of fish ovary has not received any attention so far. The extensive follicular atresia observed in the ovaries of some specimens of *X. cancila* (fig. 67) may be due to infection by such pathogens; the precise nature of this pathogen could not be determined.

The nonbreeding season is generally marked by unfavourable environmental conditions when the ovaries are reduced greatly in size and may undergo extensive atresia. In some fish species, the follicular atresia forms a conspicuous feature of postbreeding ovaries, suggesting an adaptation which temporarily suspends breeding activity during unfavourable environmental conditions. Various lines of evidence suggest that both hormonal and environmental factors regulate the initiation and also the rate of follicular atresia. But the problem of the exact mechanism(s) by which the follicles undergo atresia needs to be resolved in future studies.

The increasing use of biocides during recent years has posed a great danger to survival of fish in polluted water bodies. Therefore, the effects of various insecticides and pesticides on fish have been studied under laboratory conditions [*Lam,* 1983]. The effects of such biocides on the fish ovary form the subject of several recent studies [*Singh and Singh* 1980a–c, 1981; *Eller,* 1981; *Ram and Sathyanesan,* 1983; *Saxena and Bhatia,* 1983; *Shukla and Pandey,* 1984a, b; *Mani and Saxena,* 1985; for other references, see *Lam,* 1983]. These tonic chemical agents are observed to affect the metabolism of fish ovary and increase the incidence of follicular atresia depending upon the chemical and its dose used as well as upon the period of treatment.

In *Salmo clarki, Eller* [1981] observed reduction of ooplasm and proliferation of the follicular layer during the growth of the oocytes following chronic exposure to endrin – a chlorinated hydrocarbon. The author attributed this formation of the atypical oocytes to the inhibition of oogenesis beyond the late yolk vesicle stage and suggested that the endrin exposure interfered with the oocyte development and ovulation. *Shukla and Pandey* [1984a, b] have studied the arsenic-induced structural changes including follicular atresia during the ovarian cycle of the freshwater perch, *Colisa fasciatus.* The number of atretic follicles is increased in the ovaries of those specimens which were treated with 14.0 mg/l for 30 days. A significantly higher incidence of atresia was also observed following exposure to 5% concentration of the vegetable oil factory effluent [*Saxena and Bhatia,* 1983]. The mechanism(s) of action of biocides on fish ovary needs to be determined. These may be affecting the follicular growth (or causing follicular atresia) in the fish ovary by (1) inhibiting the secretion of gonadotrophins from the pituitary and (2) by affecting the metabolic activities of the liver as the growth of vitellogenic follicles is closely related to the synthesis of yolk (vitellogenin) by the liver (see Chapter IV). On the effect of biocides, *Singh and Singh* [1982] are of the opinion that both chlorinated and organophosphate pesticides in sublethal concentrations reduced the gonadotrophin-releasing hormone (Gn-RH)-like substance as well as gonadotrophin potency of the pituitary gland. The developmental processes of the previtellogenic follicles may be affected directly by the actions of biocides at the subcellular and molecular levels, which must be determined in future studies. *Mani and Saxena* [1985] have studied the effect of toxicologically safe concentrations of carbofuran and fenitrothion on the oocyte occurrence in the ovaries of *Channa punctatus.* Relatively more oocytes (both previtellogenic and vitellogenic) were seen under the process of atresia following fenitrothion treatment. Further, only atretic oocytes could be noticed in the ovaries of the fenitrothion-treated fish following 120 days of exposure. On the contrary, the occurrence of atretic oocytes was approximately 60% in the ovaries of the carbofuran-treated fish, while only 18% in the ovaries of the control fish, following 120 days of exposure.

The most important significance of follicular atresia during the normal course of reproduction is to limit the number of eggs that could be supported for vitellogenesis, maturation and ovulation by the female fish. But adverse environmental (including xenobiotic) factors may increase the incidence of follicular atresia as a result of stress, thus affecting the fecundity of the fish greatly. Atretic previtellogenic follicles appear to contribute interstitial cells to the ovary, which may secrete some steroid hormones needed for the growth, differentiation and maturation of normal follicles. The nature and functional significance of steroid hormones secreted by the ovarian interstitial cells must be determined in future studies.

IX. Summary

This monograph gives an up-to-date account of the cellular and molecular aspects of fish oogenesis including oocyte growth and atresia, vitellogenesis, ovum maturation and ovulation; hormonal regulation of these phases of oogenesis is also described in brief. Normal development, differentiation, structure, chemistry (including histochemistry), and function of various components of growing ovarian oocyte (or follicle) in the fish ovary are described in more detail with special emphasis on their cellular, subcellular and molecular aspects. Its various components such as the nucleus (germinal vesicle), cytoplasm and its organelles and follicular wall show distinct morphological (including ultrastructural), cytochemical and biochemical changes during previtellogenic and vitellogenic phases of oocyte growth. The vitellogenesis is characterized by the accumulation of cortical alveoli and yolk in the ooplasm.

The nuclear envelope during oocyte growth develops large folds and evaginations, consists of two parallel membranes separated by the perinuclear space, and has nuclear pores; the latter increase in number during oocyte growth. Divergent views exist about the structure of pore complex which appears to consist of nuclear pore and its associated annular material. The nuclear membranes form blebs which are detached to form part of the endoplasmic reticulum. The lampbrush chromosomes form an important site for the synthesis of RNA and protein in the nucleus (or germinal vesicle) of growing previtellogenic oocyte. Multiple nucleoli reflecting an amplification of ribosomal genes (or DNA) and thus consisting of RNA and protein originate from certain heterochromatin regions of the chromosome known as nucleolus organizers. In their number, size, shape and structure, the nucleoli vary greatly in different fish species. The nucleolus consists of ribosome-like particles 15–20 nm in diameter intermixed with fine 'granular-fibrillar' material. Multiple nucleoli at some stages of oocyte growth may develop vacuoles of variable size; the functional significance of these vacuoles is discussed. Nucleoli contribute abundant ribosomes to the ooplasm of growing oocyte. The oocytes of some teleosts show intranuclear lipid bodies which consist of phospholipids or phospholipids and triglycerides. These correspond to the ultrastructural intranuclear inclusions. Various morphological (including ultrastructural), autoradiographic and cytochemical studies have provided evidence for the transport of RNAs and proteins from the nucleus into the ooplasm. The possible mechanism for nuclear-cytoplasmic transport are diffusion of nonparticulate elements, transport of small nucleolar fragments across pores of the nuclear envelope or migration of nucleolus/nucleoli as such across the nuclear envelope or development of large nuclear blebs whereby segments of the entire nuclear envelope or its outer membrane encloses the material to be transported into the cytoplasm and may persist for sometime in the cytoplasm.

The proliferation, differentiation, structure, chemical nature and function of various ooplasmic components such as the yolk nucleus (ultrastructural elements of endoplasmic reticulum and ribosomes), free ribosomes, RNA, mitochondria, Golgi bodies, lipid bodies, multivesicular bodies and lamellar bodies are described during oocyte growth. Some of these components form a paranuclear complex in the young oocytes which is called Balbiani's vitelline body. It shows a great variability in development, organization, and composition among different species of fish. Balbiani's vitelline body continues to persist to form an important centre for the formation, multiplication and accumulation of other organelles and inclusions during oocyte growth, which are finally distributed throughout the cytoplasm before the deposition of yolk. The presence of some enzymes is also reported in fish oogenesis. But further research is needed to follow the changing pattern of stage-specific enzymes, proteins and RNAs in the nucleus and ooplasm of growing oocyte with correlative histochemical (including autoradiographic) studies.

The cortical alveoli, two types of yolk bodies named as fatty yolk and proteid yolk, are formed during fish oogenesis. Cortical alveoli containing carbohydrates and proteins are formed by the activities of endoplasmic reticulum and Golgi apparatus. Detailed chemical nature of carbohydrates varies in the cortical alveoli of different fish species, which are not yolk in a strict sense as evidenced from their function in the cortical reaction at fertilization. Cortical alveoli are not formed in the elasmobranch oocytes. Fatty yolk consisting of neutral fats (triglycerides) is generally deposited in the cytoplasm without much morphological involvement of organelles. Proteid yolk bodies (or extravesicular yolk) is formed both endogenously and exogenously. Endogenous contribution of proteid yolk formation occurs during early vitellogenesis as evidenced from the morphological association of developing yolk granules with organelles such as the mitochondria and vesicles derived from the ER and Golgi complex. Female-specific protein (vitellogenin) is synthesized and secreted by the liver under the stimulation of oestrogen either of ovarian origin during the breeding season or exogenously administered, and then transported into the growing oocytes, by the extensive micropinocytotic activities (possibly stimulated by some pituitary gonadotrophin). The vitellogenin incorporated by micropinocytosis is transferred to forming yolk spheres within the peripheral ooplasm. The proteid yolk histochemically consists of protein and lipoprotein. But the biochemical studies have demonstrated proteins (or vitellogenin), carbohydrates, glycoproteins, glycolipids, wax esters, carotenoids, etc., either in the whole ovaries or in the isolated eggs. Species variations in regard to chemical composition and molecular weight of phosphoprotein and lipovitellin derivatives of vitellogenin are described in detail.

Follicular wall consisting of zona pellucida (chorion), follicular epithelium, basal lamina and theca is developed around the growing oocyte in the fish ovary. In their development, structure and composition, these components show a great diversity in different groups of fish. Follicular epithelium functions mainly in the synthesis of proteins as evidenced by the presence of many ribosomes, elements of granular endoplasmic reticulum, well-developed Golgi complex, coated vesicles and dense bodies; possible formation of lipids is also suggested. The changing interrelationships between the follicle cell processes and oocyte microvilli during oocyte growth are described in detail. The possible mechanisms of transport of substances such as the proteins, RNA, lipids, etc., across the follicular wall are described. The zona pellucida lying between the oocyte surface and follicular epithelium shows a great diversity in its structure and chemistry in different groups of fish. It consists of carbohydrates and proteins. The synthesis of zona material is attributed either to the oocyte itself or to the follicular epithelium or to both, depending upon the structural organization of zona pellucida, as the detailed ultrastructure and chemistry of its different zones vary greatly in different fish species. Further investigations are needed to determine the origin and chemistry of zona material. The roles of follicular epithelium and theca have been discussed in relation to steroidogenesis in the follicular wall, during vitellogenesis and ovum maturation. The thecal layer during the major part of the vitellogenesis appears to be involved in the production of oestrogen precursors which are converted to 17β-oestradiol in the follicular epithelium. Isolated special thecal cells lying close to capillaries in the thecal layer possess ultrastructural and histochemical features of well-established steroid gland cells and thus are undoubtedly the steroidogenic cells.

Eggs of different teleosts develop surface structures such as adhesive filaments, hollow spines, regularly spaced bristles. Demersal eggs generally develop thick envelopes with complex lamellae. The development, structure, nature and function of adhevise devices are described. Thin zona pellucida of pelagic eggs consists of horizontal lamellae. Some teleosts with pelagic eggs also show regularly spaced bristles, protuberances, or raised hexagonal patterns. Ornate hexagonal surface structures are formed from glycoproteins secreted by the follicle cells.

The great development and complexity of the zona pellucida is responsible for the development of micropyles from specialized follicle cells in the eggs of different teleosts; the ultrastructure of micropylar cell(s) is described. The shape, structure and origin of micropyles, which facilitate fertilization, vary in different teleosts. It is suggested that the micropyles and the surface pattern of fish eggs form important characteristic criteria for identification of different teleosts.

After the oocyte is fully grown after yolk is accumulated during vitellogenesis under the influence of 17β-oestradiol produced by the follicle wall in response to pituitary gonadotrophin, it becomes ready for final maturation (or meiotic maturation) and ovulation which constitute the end-stages of oogenesis. The endocrinological aspects of oocyte final maturation and ovulation are described; these phases of oogenesis are regulated by pituitary gonadotrophin(s) and steroid hormones secreted by the follicle wall. But the cellular and molecular mechanisms involved in ovum maturation and ovulation are poorly understood and thus form a promising area for future research. The involvement of proteolytic enzymes and prostaglandins in ovulation is suggested.

The oocytes (or follicles) in different stages of growth and differentiation (i.e. previtellogenic and vitellogenic follicles) are also lost through degeneration (or atresia) from the fish ovary. Their morphological and histochemical changes are described. The intraovarian and extraovarian factors including biocides are suggested to cause oocyte atresia. The thecal cells of atretic previtellogenic follicles appear to give rise to ovarian interstitial cells which show the features of steroidogenic cells and thus synthesize steroid hormones; the precise origin, distribution and function of ovarian interstitial cells should be clarified in future. The vitellogenic follicles (or eggs), which are generally called as the corpora atretica or preovulatory corpora lutea, are degenerating structures and do not perform any endocrine function. Further studies are needed to resolve the causes and function(s) of follicular atresia.

References

Abrahim, N.; Nelly, B.; Yasheuv, A.: Oogenesis in five species of grey mullets (Teleostei, Mugilidae) from natural and landlocked habitats. Israel J. Zool. *15:* 155–172 (1966).

Adelstein, R.S.; Klee, C.B.: Smooth muscle myosin light chain kinase; in Cheung, Calcium and cell function, vol. 1, pp. 167–182 (Academic Press, New York 1980).

Adiyodi, K.G.; Adiyodi, R.G.: Reproductive biology of invertebrates, vol. I: Oogenesis, oviposition, and oosorption (Wiley, London 1983).

Ahlstrom, E.H.; Moser, H.G.: Characters useful in identification of pelagic marine fish eggs. Calcofi Rep. *21:* 121–131 (1980).

Aida, K.; Ngam, P.V.; Hibiya, T.: Physiological studies on gonadal maturation of fishes. I. Sexual difference in composition of plasma protein of ayu in relation to gonadal maturation. Bull. Jap. Soc. scient. Fish. *39:* 1091–1106 (1973).

Aketa, K.: The chemical nature and origin of cortical alveoli in the eggs of the medaka *Oryzias latipes.* Embryologia *2:* 63–66 (1954).

Amirante, G.A.: Immunochemical studies on rainbow trout (*Salmo gairdneri* R.) lipovitellin. Acta Embryol. exp. *1972:* suppl., pp. 373–383 (1972).

Anand, T.C.; Sundararaj, B.I.: Ovarian maintenance in the hypophysectomized catfish, *Heteropneustes fossilis* (Bloch), with mammalian hypophyseal and placental hormones, and gonadal and adrenocortical steroids. Gen. compar. Endocr. *22:* 154–168 (1974).

Anderson, E.: A study of the fibrillar appendages associated with the surface of eggs of the killifish, *Fundulus heteroclitus.* Anat. Rec. *154:* 308–309 (1966).

Anderson, E.: The formation of the primary envelope oocyte differentiation in teleosts. J. Cell Biol. *35:* 193–212 (1967).

Anderson, E.: Cortical alveoli formation and vitellogenesis during oocyte differentiation in the pipefish, *Syngnathus fuscus,* and the killifish, *Fundulus heteroclitus.* J. Morph. *125:* 23–60 (1968).

Anderson, E.: Comparative aspects of the ultrastructure of the female gamete. Int. Rev. Cytol. *4:* suppl., pp. 1–70 (1974).

Ando, K.: Ultracentrifugal analysis of yolk proteins in rainbow trout egg and their changes during development. Can. J. Biochem. *43:* 373–379 (1965).

Arndt, E.A.: Die Aufgaben des Kerns während der Oogenese der Teleosteer. Z. Zellforsch. *51:* 356–378 (1960a).

Arndt, E.A.: Untersuchungen über die Eihüllen von Cypriniden. Z. Zellforsch. *52:* 315–327 (1960b).

Azevedo, C.: Evolution des enveloppes ovocytaires, au cours de l'ovogenese, chez un teleosteen vivipare, *Xiphophorus helleri.* J. Microscopie *21:* 43–54 (1974).

Azevedo, C.: Intranuclear membranous inclusions in oocytes of a viviparous teleost *(Xiphophorus helleri).* J. Cell Sci. *22:* 325–334 (1976).

Azevedo, C.; Coimbra, A.: Intranuclear membranous bodies in oocytes of a viviparous teleost. 6th Eur. Congr. Electron Microsc., Jerusalem 1976, pp. 257–258.

Azevedo, C.; Coimbra, A.: Mitochondrial changes related to yolk formation in oocytes of a viviparous teleost. 9th Int. Congr. Electr. Microsc., Toronto 1978, vol. II, pp. 252–253.

Azevedo, C.; Coimbra, A.: Evolution of nucleoli in the course of oogenesis in a viviparous teleost *(Xiphophorus helleri).* Biol. Cell *38:* 43–48 (1980).

Babu, N.; Nair, N.B.: Follicular atresia in *Amblypharyngodon chakaiensis.* Z. mikrosk.-anat. Forsch. *97:* 499–504 (1983).

Ball, J.N.: Reproduction in female bony fishes. Symp. zool. Soc. Lond. *1:* 105–135 (1960).

Bara, G.: Histological and cytological change in the ovaries of mackerel *Scomber scomber* L. during the annual cycle. Istanb. Üniv. Fen Fak. Mecm. Ser. B *25:* 49–91 (1960).

Bara, G.: Histochemical localization of 3β-hydroxysteroid dehydrogenase in the ovaries of a teleost fish, *Scomber scomber* L. Gen. compar. Endocr. *5:* 284–296 (1965).

Bara, G.: Location of steroid hormone production in the ovary of *Trachurus mediterraneus.* Acta histochem. *51:* 90–101 (1974).

Barman, T.E.; Bai, N.K.; Thoai, N.V.: Studies on a herring egg phosphoprotein. Biochem. J. *90:* 555–558 (1964).

Barr, W.A.: Patterns of ovarian activity; in Barrington, Jorgensons, Perspectives in endocrinology, pp. 164–232 (Academic Press, London 1968a).

Barr, W.A.: The endocrine control of the sexual cycle in the plaice, *Pleuronectes platessa* (L.). Gen. compar. Endocr. *3:* 216–225 (1968b).

Baumeister, V.H.G.: Lampbrush chromosomes and RNA synthesis during early oogenesis of *Brachydanio rerio* (Cyprinidae, Teleostei). Z. Zellforsch. *145:* 145–150 (1973).

Baumeister, V.H.G.: Multiple Nukleolen und RNS-Synthese in der Oogenese von *Brachydanio rerio* (Teleostei). Zool. Jb. Anat. *95:* 165–185 (1976).

Beach, A.W.: Seasonal changes in the cytology of the ovary and the pituitary gland of the goldfish. Can. J. Zool. *37:* 615–625 (1959).

Beams, H.W.; Kessel, R.G.: Oocyte structure and early vitellogenesis in the trout *Salmo gairdneri.* Am. J. Anat. *136:* 105–122 (1973).

Bergink, E.W.; Wallace, R.A.: Precursor product relationship between amphibian vitellogen and the yolk proteins, lipovitellin and phosvitin. J. biol. Chem. *249:* 2897–2902 (1974).

Borchard, P.B.: Studies on the rainbow trout (*Salmo gairdneri* R.). I. Correlation between gonadal development and serum protein pattern. Ann. Biol. anim. Biochem. Biophys. *18:* 1027–1034 (1978).

Bose, K.C.; Bose, S.K.: Nucleolar extrusions in certain fishes. I. *Chela bacaila* (Ham.) 51st & 52nd Session Proc. Indian Sci. Congr., Calcutta 1964.

Braeckevelt, C.R.; McMillan, D.B.: Cyclic changes in the ovary of the brook stickleback *Eucalia inconstans* (Kirtland) J. Morph. *123:* 373–396 (1967).

Browning, H.C.: The evolutionary history of the corpus luteum. Biol. Reprod. *8:* 128–157 (1973).

Brummett, A.R.; Dumont, J.N.: Initial stages of sperm penetration into the egg of *Fundulus heteroclitus.* J. exp. Zool. *210:* 417–434 (1979).

Brummett, A.R.; Dumont, J.N.: A comparison of chorions from eggs of northern and sourthern populations of *Fundulus heteroclitus.* Copeia *1981:* 602–614 (1981).

252 Buslé, S.: Fine structure of early previtellogenic oocytes in *Mugil* (Liza) *auratus* Risso, 1810 (Teleostei, Mugilidae). Cell Tiss. Res. *207:* 123–134 (1980).

Bussmann, B.: Investigation on heat-resistant proteins in the eggs of the Baltic herring *(Clupea harengus* L.). Meeresforschung *26:* 96–108 (1977/1978).

Busson-Mabillot, S.: Evolution des enveloppes de l'ovocyte et de l'œuf chez un poisson téléostéen. J. Microscopie *18:* 23–44 (1973).

Busson-Mabillot, S.: Un type particulier de sécrétion étocribe: celui de l'appareil adhésif de l'œuf d'un poisson téléostéen. Biol. cell. *30:* 233–244 (1977).

Callan, H.G.: The nature of lampbrush chromosomes. Int. Rev. Cytol. *15:* 1–34 (1963).

Callan, H.G.: Chromosomes and nucleoli of the axolotl, *Ambystoma mexicanum.* J. Cell Sci. *1:* 85–108 (1966).

Callan, H.G.; Lloyd, L.: Lampbrush chromosomes of crested newts *Triturus cristatus.* Phil. Trans. R. Soc. *243B:* 135–219 (1960).

Caloianu-Iordachel, M.: Cytochemical and ultrastructural data on the cytoplasm of the young oocytes in the sturgeon *Huso huso* L. Revue roum. Biol. Ser. Zool. *16:* 165–169 (1971a).

Caloianu-Iordachel, M.: Ovogenesis in acipenserid fish. The morphogenesis and histochemical composition of the external membranes. Revue roum. Biol. Ser. Zool. *16:* 113–120 (1971b).

Caloianu-Iordachel, M.: Changes in cortical cytoplasm during oocyte differentiation in Acipenserid fishes. Trans. Mus. Hist. Nat. Grigore Antipa *19:* 77–82 (1978).

Cambell, C.M.: In vitro stimulation of vitellogenin incorporation into trout oocytes by salmon pituitary extracts. Ann. Biol. Anim. Biochim. Biophys. *18:* 1013–1018 (1978).

Cambell, C.M.; Idler, D.R.: Hormonal control of vitellogenesis in hypophysectomized winter flounder (*Pseudopleuronectes americanus* Walbaum). Gen. compar. Endocr. *28:* 143–150 (1976).

Cambell, C.M.; Idler, D.R.: Characterization of an estradiol-induced protein from rainbow trout serum as vitellogenin by the composition and radioimmunological cross-reactivity to ovarian yolk fraction. Biol. Reprod. *22:* 605–616 (1980).

Cambell, C.M.; Jalabert, B.: Selective protein incorporation by vitellogenic *Salmo gairdneri* oocytes in vitro. Ann. Biol. Anim. Biochim. Biophys. *19:* 429–437 (1979).

Chan, S.T.H.; Yeung, W.S.B.: Sex control and sex reversal in fish under natural conditions; in Hoar, Randall, Donaldson, Fish physiology, vol. IX: Reproduction, part B, pp. 171–222 (Academic Press, New York 1983).

Chaudhry, H.S.: Nucleolar activity in the oocytes of some marine teleostean fishes. Jl. R. microsc. Sci. *71:* 87–93 (1951).

Chaudhry, H.S.: The yolk-nucleus of Balbiani in teleostean fish. Z. Zellforsch. *37:* 455–466 (1952).

Chaudhry, H.S.: The origin and structure of the zona pellucida in the ovarian eggs of teleosts. Z. Zellforsch. *43:* 478–485 (1956).

Chaudhuri, A.; Mandal, R.K.: Nucleoli multiplication and accumulation of RNA in oocytes of catfish *(Heteropneustes fossilis).* Nucleus, Calcutta *23:* 78–83 (1980).

Cherr, G.N.; Clark, W.H., Jr.: Fine structure of the envelope and micropyles in the eggs of the white sturgeon, *Acipenser transmontanus.* Develop. Growth Different. *24:* 341–352 (1982).

Chester Jones, I.; Ball, J.N.: Ovarian pituitary relationships; in Zuckerman, The ovary, vol. 1, pp. 361–431 (Academic Press, London 1962).

Chieffi, G.: La luteogenesi nei salacei ovovivipari: Ricerche istologiche e istochimiche in *Torpedo marmorata* e *Torpedo ocellata.* Publ. Staz. Zool., Napoli *32:* 145 (1961).
Chinareva, I.D.: Changes in the follicular epithelium and formation of egg membranes in oocytes of the last year of development in *Coregonus peled* (in Russian). Arkh. Anat. Gistol. Embriol. *64:* 79–87 (1973).
Chinareva, I.D.: Formation of a pectinate sheet of the *Coregonus peled* oocyte egg membrane (based on electron microscopic data) (in Russian). Vest. Leningr. Univ. Biol. *1:* 28–34 (1975).
Chinareva, I.D.; Kirchinskaja, E.B.: Electron microscopic study of formation of egg membranes of oocytes of the last year of development in pelled (in Russian). Arkh. Anat. Gistol. Embriol. *68:* 94–99 (1975).
Chmilevskii, D.A.; Raikova, E.V.: Incorporation of ^{3}H-thymidine into early meiotic prophase oocytes of the sterlet (in Russian). Tsitologiya *1896:* 689–692 (1976).
Chopra, H.C.: A morphological and histochemical study of the oocytes of the fish, *Ophiocephalus punctatus,* with particular reference to lipids. Q. Jl microsc. Sci. *91:* 149–158 (1958a).
Chopra, H.C.: Morphological and histochemical study of oocytes of the fish, *Barbus ticto (Punctius),* with particular reference to lipids. Res. Bull. Panjab Univ. *152:* 211–221 (1958b).
Chopra, H.C.: Cytological and cytochemical study of the growing oocytes of the fish, *Boleophthalmus dussumerii.* Cellule *60:* 303–318 (1960).
Chouinard, L.A.: Sites of formation of extra nucleoli during early oocyte growth in the freshwater teleost. *Salvelinus fontinalis* (Mitcheill). Can. J. Zool. *41:* 997–1010 (1963).
Chowdhery, V.; Kala, S.; Saxena, O.P.: Morphohistological changes in the ovary of *Colisa fasciatus.* Folia morph. *27:* 182–186 (1979).
Christensen, A.K.: Leydig cells; in Hamilton, Greep, Handbook of physiology, vol. 5: Endocrinology, pp. 57–94 (American Physiological Society, Washington 1975).
Christensen, A.K.; Gillim, S.W.: The correlation of fine structure and function in steroid secreting cells with emphasis on those of the gonads; in McKerns, The gonads, pp. 415–488 (North-Holland, Amsterdam 1969).
Clérot, J.C.: Les groupements mitochondriaux des cellules germinales des poissons téléostéens cyprinides. I. Etude ultrastructurale. J. Ultrastruct. Res. *54:* 461–475 (1976).
Clérot, J.C.: Les groupements mitochondriaux des cellules germinales des poissons téléostéens cyprinides. Expl Cell Res. *120:* 237–244 (1979).
Colombo, L.; Belvedere, P.C.; Bieniarz, K.; Epter, P.: Steroid hormone biosynthesis by ovary of the carp, *Cyprinus carpio,* during wintering. Compar. Biochem. Physiol. *72B:* 367–376 (1982).
Craik, J.C.A.: Plasma levels of vitellogenin in the elasmobranch *Scyliorhinus caniculae* (lener spotted dogfish). Compar. Biochem. Physiol. *60B:* 9–16 (1978a).
Craik, J.C.A.: Kinetic studies of vitellogenin metabolism in the elasmobranch *Scyliorhinus canicula* L. Compar. Biochem. Physiol. *61A:* 355–361 (1978b).
Craik, J.C.A.: Levels of phosphoprotein in the eggs and ovaries of some fish species. Compar. Biochem. Physiol. *72B:* 507–510 (1982).
Craik, J.C.A.; Harvey, S.M.: Biochemical changes occurring during final maturation of eggs of some marine and freshwater teleosts. J. Fish Biol. *24:* 599–610 (1984).
Crim, L.W.; Idler, D.R.: Plasma gonadotrophin, estradiol and vitellogenin and gonad

phosvitin levels in relation to the seasonal reproductive cycles of female brown trout. Ann. Biol. Anim. Biochim. Biophys. *18:* 1001–1005 (1978).

Czeczuga, B.: Carotenoids in fish. XIX. Carotenoids in the eggs of *Oncorhynchus keta* (Walbaum). Hydrobiologia *63:* 43–48 (1979).

Davidson, E.H.: Gene activity in early development; 2nd ed. (Academic Press, New York 1976).

Denis, H.: Accumulation du RNA dans les oocytes des vertébrés inférieurs. Biol. cell. *28:* 87–92 (1977).

Denis, H.; Wegnez, M.: Biochemical research on oogenesis. Oocytes and liver cells of the teleost fish *Tinca tinca* contain different kinds of 5S RNA. Devl Biol. *59:* 228–236 (1977).

Dodd, J.M.: The endocrine regulation of gametogenesis and gonad maturation in fishes. Gen. compar. Endocr. *3:* suppl., pp. 675–687 (1972).

Donaldson, E.M.: Reproductive endocrinology of fishes. Am. Zool. *13:* 909–927 (1973).

Donaldson, E.M.; Hunter, G.A.: Induced final maturation, ovulation and spermiation in cultured fish; in Hoar, Randall, Donaldson, Fish physiology, vol. IX: Reproduction, part B, pp. 351–404 (Academic Press, New York 1983).

Donaldson, E.M.; Hunter, G.A.; Kraak, G. van der; Dye, H.M.: Application of LH-RH analogues to the induced final maturation and ovulation of coho salmon *(Oncorhynchus kisutch);* in Richter, Goos, Reproductive physiology of fish, pp. 177–180 (Centre for Agricultural Publishing and Documentation, Wageningen 1982).

Donato, A.; Contini, A.; Maugeri, A.; Fasulo, S.: Structural and ultrastructural aspects of the growing oocytes of *Chromis chromis* (Teleostei, Labridae). Riv. Biol. Norm. Patol. *6:* 31–66 (1980).

Drilhon, A.; Fine, J.M.: Dimorphisme séxuel dans les proteins sériques de *Salmo salar:* Étude éléctrophorétique. C.r. Séanc. Soc. Biol. *157:* 1897–1900 (1963).

Droller, M.J.; Roth, T.F.: An electron microscopic study of yolk formation during oogenesis in *Lebistes reticulatus* Guppyi. J. Cell Biol. *28:* 209–232 (1966).

Dumont, J.N.; Brummett, A.R.: The vitelline envelope, chorion and micropyle of *Fundulus heteroclitus* eggs. Gamete Res. *3:* 24–44 (1980).

Dunbar, B.S.; Wolgemuth, D.J.: Structure and function of the mammalian zona pellucida, a unique extracellular matrix; in Modern cell biology, vol. 3, pp. 77–111 (Liss, New York 1984).

Duryee, W.R.: Chromosomal physiology in relation to nuclear structure. Ann. N.Y. Acad. Sci. *50:* 920–953 (1950).

Dutt, N.H.G.; Govindan, P.; Rajasekarsetty, M.R.: Histochemical localization of alkaline and acid phosphatases in the oocytes of *Anabas scandens* (Cuvier). J. Anim. Morphol. Physiol. *22:* 151–158 (1975).

Egami, N.; Ijiri, K.-I.: Effects of irradiation on germ cells and embryonic development in teleosts. Int. Rev. Cytol. *59:* 195–244 (1979).

Eggert, B.: Entwicklung und Bau der Eier von *Salarius flavoumbrinus* Rupp. Zool. Anz. *83:* 241–253 (1929).

Eller, L.L.: Histopathological lesions in cut-throat trout, *Salmo salmo clarki,* exposed chronically to the insecticide endrin. Am. J. Path. *64:* 321–336 (1981).

Emel'Yanova, N.G.: Electron microscopic study of the development of egg membranes and follicle of the silver carp *Hypophthalmichthys molitrix.* Vopr. Ikhtiol. *19:* 302–312 (1979).

Emmersen, B.K.; Emmersen, J.: Protein, RNA and DNA metabolism in relation to ovarian vitellogenic growth in the flounder *Platichthys flesus* (L). Compar. Biochem. Physiol. *55B:* 315–321 (1976).

Emmersen, B.K.; Petersen, J.: Natural occurrence, and experimental induction by estradiol-17β of a lipophosphoprotein (vitellogenin) in flounder (*Platichthys flesus* L.). Compar. Biochem. Physiol. *54B:* 443–446 (1976).

Epler, P.: Effects of steroid and gonadotropic hormones on the maturation of carp ovaries. V. Ovulation, fertilization, and embryonal development of carp oocytes in vitro. Pol. Archs Hydrobiol. *28:* 119–125 (1981).

Erhardt, H.: Licht- und elektronenmikroskopische Untersuchungen an den Eihüllen des marinen Teleosteers *Lutjanus sgnagris.* Helgoländer wiss. Meeresunters. *28:* 90–105 (1976).

Erhardt, H.: Elektronenmikroskopische Untersuchungen an den Eihüllen von *Lutianus analis* (Cuvier & Valenciennes, 1828) (Lutjanidae, Perciformis, Pisces). Biol. Zbl. *97:* 181–187 (1978).

Erhardt, H.; Götting, M.J.: Light and electron microscopic studies in egg cells and envelopes of *Platypoecilus maculatus* (in German). Cytobiologie *2:* 429–440 (1970).

Espey, L.L.: Ovulation; in Jones, The vertebrate ovary, pp. 503–532 (Plenum Press, New York 1978).

Ficq, A.: Transition du stade pachytène au stade diplotène de la méiose chez *Xenopus laevis.* Etude cytochimique et autoradiographique. J. Microsc. Biol. cell. *25:* 7–16 (1976).

Fisher, K.C.: The formation and properties of the external membrane of the trout egg. Trans. R. Soc. Can. *1:* 323–332 (1963).

Flegler, C.: Electron microscopic studies on the development of the chorion of the viviparous teleost *Dermogenys pusillus* (Hemirhamphidae). Cell Tiss. Res. *179:* 155–279 (1977).

Fletcher, T.C.; White, A.; Youngson, A.; Pusztai, A.; Baldo, B.A.: Characterization of C-reactive protein from the eggs of the marine teleost *Cyclopterus lumpus.* Biochim. biophys. Acta *671:* 44–49 (1981).

Flügel, H.: Electron microscopic investigations on the fine structure of the follicular cells and the zona radiata of trout oocytes during and after ovulation. Naturwissenschaften *51:* 564–565 (1964).

Flügel, H.: Licht- und elektronenmikroskopische Untersuchungen an Oozyten und Eiern einiger Knochenfische. Z. Zellforsch. *83:* 82–116 (1967a).

Flügel, H.: Elektronenmikroskopische Untersuchungen an den Hüllen der Oozyten und Eiern des Flussbarsches *Perca fluviatilis.* Z. Zellforsch. *77:* 245–256 (1967b).

Flügel, H.: Zur Funktion des Golgi-Apparates in den Follikelzellen von *Coregonus albala* L. Cytobiologie *4:* 450–459 (1970).

Fontaine, M.: Hormones and the control of reproduction in aquaculture. J. Fish. Res. Bd Can. *33:* 922–939 (1976).

Fontaine, Y.A.: Hormones in fishes; in Malins, Sargent, Biochemical and biophysical perspectives in marine biology, vol. 2, pp. 139–212 (Academic Press, New York 1975).

Fostier, A.; Jalabert, B.: Physiological basis of practical means to induce ovulation in fish; in Richter, Goos, Reproductive physiology of fish, pp. 164–173 (Centre for Agricultural Publishing and Documentation, Wageningen 1982).

Fostier, A.; Jalabert, B.; Bilard, R.; Breton, B.; Zona, Y.: The gonadal steroids; in Hoar, Randall, Donaldson, Fish physiology, vol. IX: Reproduction, part A, pp. 227–372 (Academic Press, New York 1983).

Gabaeva, N.S.: Characteristics of the follicular epithelium in the fishes with demersal roe. Concerning the classification of unilaminate monomorphous follicular epithelium of the vertebrates (in Russian). Arkh. Anat. Gistol. Embriol. *73:* 48–55 (1977).

Gabaeva, N.S.; Boglazova, E.K.: The morphodynamics and proliferative activity of the follicular epithelium in the oogenesis of the sword-tailed minnow *Xiphophorus helleri* (in Russian). Biol. Nauki, Moscow *20:* 59–65 (1977).

Gabaeva, N.S.; Ermolina, N.O.: On changes on the follicular epithelium of ovum membranes during oogenesis of the sheat-fish *Arius thalassinus* (in Russian). Arkh. Anat. Gistol. Embriol. *63:* 97–106 (1972).

Ginsburg, A.S.; Fertilization in fishes and the problem of polyspermy (in Russian). Akad. Nauk. SSR (1968).

Goetz, F.W.: Hormonal control of oocyte final maturation and ovulation in fishes; in Hoar, Randall, Donaldson, Fish physiology, vol. IX: Reproduction, part B, pp. 117–170 (Academic Press, New York 1983).

Goetz, F.W.; Bergman, H.L.: The effects of steroids on final maturation and ovulation of oocytes from brook trout *(Salvelinus fontinalis)* and yellow perch *(Perca flavescens).* Biol. Reprod. *18:* 293–298 (1978).

Goetz, F.W.; Smith, D.C.; Krickl, S.P.: The effects of prostaglandins, phosphodiesterase inhibitors and cyclic AMP on ovulation of brook trout *(Salvelinus fontinalis)* oocytes. Gen. compar. Endocr. *48:* 154–160 (1982).

Goetz, F.W.; Theofan, G.: In vitro stimulation of germinal vesicle breakdown and ovulation of yellow perch *(Perca flavescens)* oocytes. Effects of 17α-hydroxy-20β-dihydroprogesterone and prostaglandins. Gen. compar. Endocr. *37:* 273–285 (1979).

Gopal Dutt, N.H.: The yolk nucleus in the oocytes of *Anabas scandens.* Q. Jl microsc. Sci. *105:* 349–352 (1964).

Gopal Dutt, N.H.; Govindan, P.: The localization of lipids during the oogenesis in *Anabas scandens* (Cuvier). Z. mikrosk.-anat. Forsch. *74:* 179–192 (1967).

Gopal Dutt, N.H.; Govindan, P.: Localization of polysaccharides, -SH and -SS groups in the oocytes of *Anabas scandens* (Cuivier). Z. mikrosk.-anat. Forsch. *80:* 399–418 (1969).

Goswami, S.V.; Sundararaj, B.I.: Temporal effects of ovine luteinizing hormone and desoxycorticosterone acetate on maturation and ovulation of oocytes of the catfish *Heteropneustes fossilis* (Bloch): an in vivo and in vitro study. J. exp. Zool. *178:* 457–466 (1971).

Götting, K.J.: Die Feinstruktur der Hüllschichten reifender Oocyten von *Agonus cataphractus* L. (Teleostei, Asonidae). Z. Zellforsch. *66:* 405–414 (1965).

Götting, K.J.: Zur Feinstruktur der Oocyten mariner Teleosteer. Helgoländer wiss. Meeresunters. *13:* 118–170 (1966).

Götting, K.J.: Der Follikel und die peripheren Strukturen der Oocyten der Teleosteer und Amphibien. Z. Zellforsch. *79:* 481–491 (1967).

Götting, K.J.: Zur Darstellung der Ultrastruktur des Teleosteer-Follikels mittels der Gefrierätztechnik. Micron *1:* 356–372 (1970).

Götting, K.J.: Oocyte ultrastructure of oviparous and ovoviviparous teleosts as revealed by freeze etching. Int. Congr. Electron Microsc. Canberra *8:* 668–669 (1974).

Götting, K.J.: Fortpflanzung und Oocyten-Entwicklung bei der Aalmutter *(Zoarces viviparus)* (Pisces, Osteichthyes). Helgoländer wiss. Meeresunters. *28:* 71–89 (1976).
Graham, Y.; Kagawa, H.; Nagahama, Y.: Evidence for a decrease in aromatase activity in the ovarian granulosa cells of amago salmon *(Oncorhynchus rhodurus)* associated with final oocyte maturation. Biol. Reprod. *29:* 310–315 (1983).
Grodzinski, Z.: The yolk of bitterling *Rhodeus amarus* Bl. Folia morph. *13:* 13–36 (1954).
Grodzinski, Z.: Zohko nizszych kregowcow. Przegl. Zool. *17:* 159–171 (1973).
Gupta, N.N.; Yamamoto, K.: Electron microscope study on the fine structural changes in the oocytes of the goldfish, *Carassius auratus,* during yolk formation stage. Bull. Fac. Fish. Hokkaido Univ. *22:* 187–206 (1972).
Guraya, S.S.: Histochemical studies on the yolk nucleus in the fish oogenesis. Z. Zellforsch. *60:* 659–666 (1963).
Guraya, S.S.: A comparative histochemical study of fish *(Channa marulius)* and amphibian *(Bufo stomaticus)* oogenesis. Z. Zellforsch. *65:* 662–700 (1965).
Guraya, S.S.: Morphology, histochemistry and biochemistry of human ovarian compartments and steroid hormone synthesis. Physiol. Rev. *5:* 785–807 (1971).
Guraya, S.S.: Histochemical observations on the interstitial gland cells of dogfish ovary. Gen. compar. Endocr. *18:* 409–412 (1972).
Guraya, S.S.: Follicular atresia; in Proceedings of the Golden Jubilee Symposium on Perspectives in Reproductive Physiology of the Female, Delhi University. Proc. Indian natn. Sci. Acad. *39B:* 311–332 (1973).
Guraya, S.S.: Gonadotrophins and functions of granulosal and thecal cells in vivo and in vitro; in Moudgal, Gonadotropins and Gonadal function, pp. 220–236 (Academic Press, New York 1974).
Guraya, S.S.: Recent advances in the morphology, histochemistry and biochemistry of steroid-synthesizing cellular sites in the nonmammalian vertebrate ovary. Int. Rev. Cytol. *44:* 365–409 (1976a).
Guraya, S.S.: Recent advances in the morphology, histochemistry and biochemistry of steroid-synthesizing cellular sites in the non-mammalian vertebrate testis; in Bourne, Danielli, International review of cytology, vol. 47, pp. 99–136 (Academic Press, New York 1976b).
Guraya, S.S.: Maturation of the follicular wall of nonmammalian vertebrates; in Jones, The vertebrate ovary, pp. 261–329 (Plenum Press, New York 1978).
Guraya, S.S.: Recent advances in the morphology and histochemistry of steroid-synthesizing cellular sites in the gonads of fish. Proc. Indian natn. Sci. Acad. *45B:* 452–461 (1979a).
Guraya, S.S.: Recent advances in the morphology, cytochemistry and function of Balbiani's vitelline body in animal oocytes. Int. Rev. Cytol. *59:* 249–321 (1979b).
Guraya, S.S.: Recent progress in the structure, origin, composition, and function of cortical granules in animal egg. Int. Rev. Cytol. *78:* 257–360 (1982).
Guraya, S.S.: Biology of ovarian follicles in mammals (Springer, Heidelberg 1985).
Guraya, S.S.; Kaur, S.: Morphology of the postovulatory follicle (or corpus luteum) of teleost (*Cyprinus carpio* L.) ovary. Zool. Beitr. *25:* 381–390 (1979).
Guraya, S.S.; Kaur, S.: Cellular sites of steroid synthesis in the oviparous teleost fish (*Cyprinus carpio* L.): a histochemical study. Proc. Indian Acad. Sci. (Anim. Sci.) *91:* 587–597 (1982).

Guraya, S.S.; Kaur, S.; Saxena, P.K.: Morphology of ovarian changes during reproductive cycle of fish, *Mystus tengara* (Ham.). Acta anat. *91:* 222–260 (1975).

Guraya S.S.; Motta, P.: Interstitial cells and related structures; in Hafez, Motta, Biology of the ovary, chapt. 5, pp. 69–95 (Nijhoff, The Hague 1980).

Guraya, S.S.; Toor, H.S.; Kumar, S.: Morphology of ovarian changes during the reproductive cycle of the *Cyprinus carpio* communis (Linn.). Zool. Beitr. *23:* 405–437 (1977).

Hagenmaier, H.E.: The hatching process in fish embryos. III. The structure, polysaccharide and protein cytochemistry of the chorion of the trout egg, *Salmo gairdneri* (Rich.). Acta histochem. *47:* 61–69 (1973).

Hagstrom, B.E.; Lönning, S.: Electron microscopic studies of unfertilized and fertilized eggs from marine teleosts. Sarasia *33:* 73–80 (1968).

Hara, A.: Electrophoretical and immunological studies on fish serum proteins. Bull. Jap. Soc. scient. Fish. *41:* 105–113 (1975).

Hara, A.: Iron-binding activity of female-specific serum proteins of rainbow trout *(Salmo gairdneri)* and chum salmon *(Oncorhynchus keta).* Biochim. biophys. Acta *427:* 549–557 (1976).

Hara, A.; Hirai, H.: Comparative studies on immunochemical properties of female specific serum protein and egg yolk proteins in rainbow trout *(Salmo gairdneri).* Compar. Biochem. Physiol. *59B:* 339–343 (1978).

Hara, A.; Yamauchi, K.; Hirai, H.: Studies on female-specific serum protein (vitellogenin) and egg yolk protein in Japanese eel *(Anguilla japonica).* Compar. Biochem. Physiol. *65B:* 315–320 (1980).

Hart, N.H.; Donovan, M.: Fine structure of the chorion and site of sperm entry in the egg of *Brachydanio rerio.* J. exp. Zool. *227:* 277–296 (1983).

Harvey, B.J.; Hoar, W.S.: The theory and practice of induced breeding in fish (IDRC, Ottawa 1979).

Heesen, D. te: Immunologische Untersuchungen an exo- und endogenen Dotterproteinen von *Brachydanio rerio* (Teleostei, Cyprinidae) und verwandten Arten. Zool. Jb. Anat. *97:* 566–582 (1977).

Heesen, D. te; Engel, W.: Electrophoretic studies on vitellogenesis in *Brachydanio rerio* (Cyprinidae, Teleostei). Wilhelm Roux Arch. devl Biol. *173:* 46–59 (1973).

Hess, O.; Meyer, G.T.: Chromosomal differentiation of the lampbrush type formed by the Y chromosome in *Drosophila hydei* and *D. neohydei.* J. Cell Biol. *16:* 527–539 (1963).

Hickey, E.D.; Wallace, R.A.: Study of the vitellogenin protein in the serum of estrogen-treated *Ictalurus nebulosus.* Biol. Bull. *147:* 481 (1974).

Hinegardner, R.: Evolution of cellular DNA content in teleost fish species in nature. Am. Nat. *102:* 517–523 (1968).

Hirose, K.: The ultrastructure of the ovarian follicle of medaka, *Oryzias latipes.* Z. Zellforsch. *123:* 316–329 (1972).

Hirose, K.: Endocrine control of ovulation in Medaka *(Oryzias latipes)* and Aya *(Plecoglossus altivelis).* J. Fish. Res. Bd Can. *33:* 989–994 (1976).

Hirose, K.; Adachi, S.; Nagahama, Y.: Changes in plasma steroid hormone levels during sexual maturation in the Ayu *Plecoglossus altivelis.* Bull. Jap. Soc. scient. Fish. *51:* 399–403 (1985).

Hirose, K.; Machida, Y.; Donaldson, E.M.: Induction of ovulation in the Japanese flounder *(Limanda yokohamae)* with human chorionic gonadotropin and salmon gonadotrophin. Bull. Jap. Soc. scient. Fish. *42:* 13–20 (1976).

Hisaoka, K.K.; Firlit, C.F.: The localization of nucleic acids of the zebra fish. Am. J. Anat. *110:* 203–216 (1962).

Hisaw, F.L., Jr.; Hisaw, F.L.: The corpora lutea of elasmobranch fishes. Anat. Rec. *135:* 269–277 (1960).

Ho, F.C.W.; Vanstone, W.E.: Effect of estradiol monobenzoate on some serum constituents of maturing sockeye salmon *(Oncorhynchus nerka).* J. Fish. Res. Bd Can. *18:* 859–864 (1961).

Hoar, W.S.: Comparative physiology: hormones and reproduction in fishes. A. Rev. Physiol. *27:* 51–70 (1965).

Hoar, W.S.; Ngahama, Y.: The cellular sources of sex steroids in teleost gonads. Ann. Biol. Anim. Biochim. Biophys. *18:* 893–898 (1978).

Hori, S.H.; Kodama, T.; Tanahashi, K.: Induction of vitellogenin synthesis in goldfish by massive doses of androgens. Gen. compar. Endocr. *37:* 306–320 (1979).

Hosokawa, K.: Scanning electron microscopic observations of the micropyle in *Oryzias latipes.* Jap. J. Ichthyol. *26:* 94–99 (1979).

Hosokawa, K.; Fusimi, T.; Matsusato, T.: Electron microscopic observation of the chorion and micropyle apparatus of the porgy, *Pagrus major.* Jap. J. Ichthyol. *27:* 339–343 (1981).

Hunter, G.A.; Donaldson, E.M.: Hormonal sex control and its application to fish culture; in Hoar, Randall, Donaldson, Fish physiology, vol. IX: Reproduction, part B, pp. 223–304 (Academic Press, New York 1983).

Hurk, R. van den; Peute, J.: Cyclic changes in the ovary of the rainbow trout, *Salmo gairdneri,* with special reference to sites of steroidogenesis. Cell Tiss. Res. *199:* 289–306 (1979).

Hurk, R. van den; Richter, C.J.J.: Histochemical evidence for granulosa steroids in follicle maturation in the African catfish *Clarias lazera.* Cell Tiss. Res. *211:* 345–348 (1980).

Hurley, D.A.; Fisher, K.C.: The structure and development of the external membrane in young eggs of the brook trout, *Salvelinus fontinalis* (Mitcheill). Can. J. Zool. *44:* 173–189 (1966).

Idler, D.R.; Campbell, C.M.: Gonadotropin stimulation of estrogen and yolk precursor synthesis in juvenile rainbow trout. Gen. compar. Endocr. *41:* 384–391 (1980).

Idler, D.R.; Hwang, S.J.; Crim, L.W.: Quantification of vitellogenin in Atlantic salmon *(Salmo salar)* plasma by radioimmunoassay. J. Fish. Res. Bd Can. *36:* 574–578 (1979).

Idler, D.R.; Hwang, S.J.; Crim, L.W.; Reddin, D.: Determination of sexual maturational stages of Atlantic salmon *(Salmo salar)* captured at sea. Can. J. Fish. aquat. Sci. *38:* 405–413 (1981).

Idler, D.R.; Ng, T.B.: Studies on two types of gonadotropins from both salmon and carp pituitaries. Gen. compar. Endocr. *38:* 421–440 (1979).

Iino, A.; Inoue, T.: Fine structure of the zona pellucida and the attaching filaments in oocytes from the teleosts. Okajimas Folia anat. jap. *58:* 1037–1052 (1982).

Inoue, S.; Iwasaki, A.: Isolation of a novel glycoprotein for the eggs of rainbow trout: occurrence of disialosyl groups on all carbohydrate chains. Biochem. biophys. Res. Commun. *83:* 1018–1023 (1978).

Inoue, S.; Iwasaki, A.: Sialoglycoprotein from the eggs of Pacific herring *(Clupea pallasii).* Eur. J. Biochem. *111:* 131–138 (1980).

Inoue, S.; Kaneda-Hyashi, T.; Sugiyama, H.; Ando, T.: Studies on phosphoproteins from fish eggs. I. Isolation and characterization of a phosphoprotein from the eggs of Pacific herring. J. Biochem. *69:* 1003–1011 (1971).

Ionescu, V.M.: Incorporation of tritium-labeled thymidine in the oocytes of *Carassius auratus gibelio.* Revue roum Biol. Ser. Biol. Anim. *25:* 121–124 (1980).

Ivankov, V.N.; Kurdyayeva, V.P.: Systematic differences and the ecological importance of the membranes in fish eggs. J. Ichthyol. *13:* 864–873 (1973).

Iwamatsu, T.: On fertilizability of pre-ovulation eggs in the medaka, *Oryzias latipes.* Embryologia *8:* 327–336 (1965).

Iwamatsu, T.: Studies on oocyte maturation of the medaka *Oryzias latipes.* VIII. Role of follicular constituents in gonadotropin and steroid-induced maturation of oocytes in vitro. J. exp. Zool. *211:* 231–239 (1980).

Iwamatsu, T.; Katoh, T.: Maturation in vitro of oocytes of the loach, *Misgurnus anguillicaudatus* by steroid hormones and gonadotropins. Annotnes zool. jap. *51:* 79–89 (1978).

Iwamatsu, T.; Keino, H.: Scanning electron microscopic study on the surface change of eggs of the teleost, *Oryzias latipes,* at the time of fertilization. Develop. Growth Different. *20:* 237–250 (1978).

Iwamatsu, T.; Ohta, T.: Breakdown of the cortical alveoli of medaka eggs at the time of fertilization with a particular reference to the possible role of spherical bodies in the alveoli. Wilhelm Roux Arch. devl Biol. *180:* 297–309 (1976).

Iwamatsu, T.; Ohta, T.: Fine structure of loach oocytes during maturation in vitro. Develop. Growth Different. *19:* 213–226 (1977).

Iwamatsu, T.; Ohta, T.: Scanning electron microscopic observations on sperm penetration in teleostean fish J. exp. Zool. *218:* 261–277 (1981a).

Iwamatsu, T.; Ohta, T.: On a relationship between oocyte and follicle cells around the time of ovulation in the medaka, *Oryzias latipes.* Annotnes zool. jap. *54:* 17–29 (1981b).

Iwamatsu, T.; Ohta, T.; Nakayama, N.; Shoji, H.: Studies of oocyte maturation of the medaka, *Oryzias latipes.* Annotnes zool. jap. *49:* 28–37 (1976).

Iwasaki, Y.: Histochemical detection of Δ^5-3β-hydroxysteroid dehydrogenase in the ovary of medaka, *Oryzias latipes* during annual reproductive cycle. Bull. Fac. Fish. Hokkaido Univ. *23:* 177–184 (1973).

Iyengar, R.; Schlenk, H.: Wax esters of mullet *(Mugil cephalus)* roe oil. Biochemistry *6:* 396–402 (1967).

Jalabert, B.: Modulation par différents stéroides non maturants de l'efficacité de la 17α-hydroxy-20β-dihydroprogestérone ou d'un extrait gonadotrope sur la maturation intra-folliculaire in vitro des ovocytes de la truite arc-enciel, *Salmo gairdnerii.* C.r. hebd. Séanc. Acad. Sci., Paris *281D:* 811–814 (1975).

Jalabert, B.: In vitro oocyte maturation and ovulation in rainbow trout *(Salmo gairdneri),* northern pike *(Esox lucius)* and goldfish *(Carassius auratus).* J. Fish. Res. Bd Can. *33:* 974–988 (1976).

Jalabert, B.: Production of fertilizable oocytes from follicles of rainbow trout *(Salmo gairdnerii)* following in vitro maturation and ovulation. Ann. Biol. Anim. Biochim. Biophys. *18:* 461–470 (1978).

Jalabert, B.; Breton, B.; Fostier, A.: Precocious induction of oocyte maturation and ovulation in rainbow trout *(Salmo gairdneri):* problems when using 17α-hydroxy-20β-dihydroprogesterone. Ann. Biol. Anim. Biochim. Biophys. *18:* 977–984 (1978).

Jalabert, B.; Fostier, A.: The follicular sensitivity in vitro to maturation-inducing hormones in rainbow trout, *Salmo gairdneri:* role of oestradiol-17β. Aquaculture *43:* 1–11 (1984a).

Jalabert, B.; Fostier, A.: The modulatory effect in vitro of oestradiol-17β, testosterone or cortisol on the output of 17α-hydroxy-20β-dihydroprogesterone by rainbow trout *(Salmo gairdneri)* ovarian follicles stimulated by the maturational gonadotropin s-GtH. Reprod. Nutr. Dev. *24:* 127–136 (1984b).

Jalabert, B.; Szöllösi, D.: In vitro ovulation of trout oocytes: effects of prostaglandins on smooth muscle-like cells of the theca. Prostaglandins *9:* 765–778 (1975).

Jared, D.W.; Wallace, R.A.: Comparative chromatography of the yolk proteins of teleosts. Compar. Biochem. Physiol. *24:* 437–443 (1968).

Jollie, W.P.; Jollie, L.C.: The fine structure of the ovarian follicle of the ovoviviparous Poeciliid fish, *Lebistes reticulatus.* I. Maturation of follicular epithelium. J. Morph. *114:* 479–502 (1964).

Kagawa, H.; Nagahama, Y.: In vitro effects of prostaglandins on ovulation in goldfish *Carassius auratus.* Bull. Jap. Soc. scient. Fish. *47:* 1119–1121 (1981).

Kagawa, H.; Takano, K.: Ultrastructure and histochemistry of granulosa cells of pre- and post-ovulatory follicles in the ovary of the medaka *Oryzias latipes.* Bull. Fac. Fish. Hokkaido Univ. *30:* 191–204 (1979).

Kagawa, H.; Takano, K.; Nagahama, Y.: Correlation of plasma estradiol-17β and progesterone levels with ultrastructure and histochemistry of ovarian follicles in the white-spotted char, *Salvelinus leucomaenis.* Cell Tiss. Res. *218:* 315–329 (1981).

Kagawa, H.; Young, G.; Adachi, S.; Nagahama, Y.: Estradiol-17β production in amago salmon *(Oncorhynchus rhodurus)* ovarian follicles: role of the thecal and granulosa cells. Gen. compar. Endocr. *47:* 440–448 (1982a).

Kagawa, H.; Young, G.; Nagahama, Y.: Estradiol-17β production in isolated amago salmon *(Oncorhynchus rhodurus)* ovarian follicles and its stimulation by gonadotropins. Gen. compar. Endocr. *47:* 361–365 (1982b).

Kagawa, H.; Young, G.; Nagahama, Y.: Relationship between seasonal plasma estradiol-17β and testosterone levels and in vitro production by ovarian follicles of amago salmon *(Oncorhynchus rhodurus).* Biol. Reprod. *29:* 301–309 (1983).

Kagawa, H.; Young, G.; Nagahama, Y.: In vitro estradiol-17β and testosterone production by ovarian follicles of the goldfish, *Carassius auratus.* Gen. compar. Endocr. *54:* 139–143 (1984).

Kaitaranta, J.K.; Ackman, R.G.: Total lipids and lipid classes of fish roe. Compar. Biochem. Physiol. *69B:* 725–729 (1981).

Kapoor, C.: Polynuclear oocyte in *Puntius ticto* (Hamilton). J. Anim. Morphol. Physiol. *22:* 221 (1976a).

Kapoor, C.: Studies on the oogonial cycle in *Puntius ticto* (Ham.). II. Formation of mature follicles. Z. mikrosk.-Anat. Forsch. *9:* 1049–1055 (1976b).

Kapoor, C.: Nucleolar extrusion in the oocytes of *Puntius ticto* (Ham.). Zool. Beitr. *23:* 221–226 (1977).

Kapur, K.: Effects of indomethacin on ovulation and spawning in exotic fish *Cyprinus carpio.* Indian J. exp. Biol. *17:* 517–519 (1979).

Kemp, N.E.; Allen, M.D.: Electron microscopic observations on the development of the chorion of *Fundulus*. Biol. Bull. mar. biol. Lab., Woods Hole *11:* 293 (1956).

Khanna, S.S.; Pant, M.C.: Seasonal changes in the ovary of a sisorid catfish *Glyptosternum poetinopterum*. Copeia *1:* 83–88 (1967).

Khoo, K.H.: The corpus luteum of goldfish *(Carassius auratus* L.) and its function. Can. J. Zool. *53:* 1306–1323 (1975).

Khoo, K.H.; The histochemistry and endocrine control of vitellogenesis in goldfish ovaries. Can. J. Zool. *57:* 617–626 (1979).

Kim, E.D.; Zhukinskii, V.M.: Amino acid variability in bream roe during maturation and overmaturation. Dopov. Akad. Nauk. ukr. RSR (in Russian). *1978:* 272–277 (1978).

Kondrat'Ev, A.K.: Cytoecological studies on the oogenesis of *Acipenser ruthenus*. Seasonal dynamics of distribution of RNA-rich cytoplasmic structures in developing oocytes during previtellogenesis (in Russian). Tsitologiya *21:* 1397 (1979).

Konopacka, B.: Recherches histochimiques sur le développement des poissons. I. La vitellogenèse chez le goujon *(Gobio fluviatilis)* et la carpe *(Cyprinus carpio)*. Bull. int. Acad. pol. Sci. Lett. Ser. B. *1935:* 163–182 (1935).

Konovalov, Y.D.: A study of some functional groups of proteins in egg cells of the black sea roach and carp-bream eggs of different degrees of maturity (in Russian). Ontogenez *10:* 359–364 (1979).

Konovalov, Y.D.; Zhukinskii, V.M.: Proteins and their reactive groups in roe and sperm of roach and carp-bream (in Russian). Ukr. Biokhim. Zh. *51:* 7–9 (1979).

Korfsmeier, K.H.: Zur Genese des Dottersystems in der Oocyte von *Brachydanio rerio*. Autoradiographische Untersuchungen. Z. Zellforsch. *71:* 283–296 (1966).

Korsgaard, B.; Petersen, I.: Vitellogenin, lipid and carbohydrate metabolism during vitellogenesis and pregnancy and after hormonal induction in the blenny *Zoarces viviparus* (L.). Compar. Biochem. Physiol. *63B:* 245–251 (1979).

Kosmath, I.; Patzner, R.A.; Adam, H.: The structure of the micropyles of the eggs of *Myxine glutinosa* and *Eptatretus burgeri* (Cyclostomata). Zool. Anz. *206:* 273–278 (1981).

Kosmath, I.; Patzner, R.A.; Adam, H.: Regression in the ovary of *Myxine glutinosa* (Cyclostomata). 2. Electron microscopic examinations of atretic follicles. Z. mikrosk.-anat. Forsch. *97:* 667–674 (1983a).

Kosmath, I.; Patzner, R.A.; Adam, H.: Regression in the ovary of *Myxine glutinosa* (Cyclostomata). 1. Degeneration of yolk platelets in the atretic follicles. Zool. Anz. *211:* 273–276 (1983b).

Kosmath, I.; Patzner, R.A.; Adam, H.: Regression in the ovary of *Myxine glutinosa*. 4. Histochemical examination of atretic follicles. Z. mikrosk.-anat. Forsch. *97:* 941–947 (1983c).

Kraak, G. van der: An introduction to gonadotropin receptor studies in fish; in Hoar, Randall, Donaldson, Fish physiology, vol. IX: Reproduction, part A, pp. 405–441 (Academic Press, New York 1983).

Kraak, G. van der; Hoa-Ren Lin; Donaldson, E.M.; Dye, H.M.; Hunter, G.A.: Effects of LHRH and (des-10-glycine, 6-*D*-alanine) LHRH-ethylamide on plasma gonadotropin levels and oocyte maturation in adult female coho salmon *(Oncorhynchus kisutch)*. Gen. compar. Endocr. *49:* 470–476 (1983).

Kraft, A.V.; Peters, H.M.: Vergleichende Studien über die Oogenese in der *Gattung tilpaia* (Cichlidae, Teleostei). Z. Zellforsch. *61:* 434–485 (1963).

Krajhanzl, A.; Nosek, J.; Habrova, V.; Kocourek, J.: An immunofluorescence study on the occurrence of endogenous lectins in the differentiating oocytes of silver carp *(Hypophthalmichthys molitrix* Valenc.) and tench *(Tinca tinca* L.). Histochem. J. *16:* 432–434 (1984a).

Krajhanzl, A.; Nosek, J.; Monsigny, M.; Kocourek, J.: Direct visualization of endogenous lectins in fish oocytes by glycosylated fluorescent cytochemical markers. Histochem. J. *16:* 426–428 (1984b).

Krauel, K.K.; Ridgeway, G.J.: Immunoelectrophoretic studies of red salmon *(Oncorhynchus nerka)* serum. Int. Archs Allergy appl. Immun. *23:* 246–253 (1963).

Krishnamurthy, C.G.; Rajasekharasetty, M.R.; Gopal Dutt, N.H.: Oocytes of *Rusbora daniconius* (Ham.); chemical composition and fate of yolk vesicles. Indian J. exp. Biol. *10:* 29–33 (1972).

Kuchnow, K.P.; Scott, J.R.: Ultrastructure of the chorion and its micropyle apparatus on the mature *Fundulus heteroclitus* (Walbaum) ovum. J. Fish Biol. *10:* 197–201 (1977).

Kudo, S.: Sperm penetration and the formation of a fertilization cone in the common carp egg. Develop. Growth Different *22:* 403–414 (1980).

Kumari, S.D.R.; Nair, N.B.: Oogenesis in the hill stream loach *Noemacheilus triangularis.* Zool. Anz. *203:* 259–271 (1979).

Kumari, S.D.R.; Padmanabhan, G.: Oogenesis in the pearl-spot *Etroplus suratensis* (Bl.). Zool. Anz. *196:* 133–143 (1976).

Kuo, C.-M.: Induced breeding of grey mullet, *Mugil cephalus;* in Richter, Goos, Reproductive physiology of fish, pp. 181–184 (Centre for Agricultural Publishing and Documentation, Wageningen 1982).

Kuo, C.-M.; Nash, C.E.; Shedadeh, Z.H.: A procedural guide to induce spawning in grey mullet *(Mugil cephalus* L.). Aquaculture *2:* 1–14 (1974).

Kusa, M.: On some properties of cortical alveoli in the egg of the stickleback. Annotnes zool. jap. *26:* 138–144 (1953).

Kusa, M.: Studies on cortical alveoli in some teleostean eggs. Embryologia *3:* 105–129 (1956).

Laale, H.W.: The perivitelline space and egg envelopes of bony fishes: a review. Copeia *1980:* 210–226 (1980).

Lal, B.. Morphological and cytochemical studies on the oocytes of *Cirrhina mrigala* (Hamilton) with particular reference to lipids. Proc. natn. Inst. Sci. India *29B:* 585–601 (1963).

Lam, T.J.: Environmental influences on gonadal activity in fish; in Hoar, Randall, Donaldson, Fish physiology, vol. IX: Reproduction, part B, pp. 65–116 (Academic Press, New York 1983).

Lam, T.J.; Pandey, S.; Nagahama, Y.; Hoar, W.S.: Endocrine control of oogenesis, ovulation and oviposition in goldfish; in Gaillara, Hoar, Comparative endocrinology, pp. 55–64 (Elsevier/North-Holland Medical Press, Amsterdam 1978).

Lambert, J.G.D.: Location of hormone production in the ovary of the guppy, *Poecilia reticulata.* Experientia *22:* 476 (1966).

Lambert, J.G.D.: The ovary of the guppy *Poecilia reticulata.* The granulosa cells as the sites of steroid biosynthesis. Gen. compar. Endocr. *15:* 464–476 (1970a).

Lambert, J.G.D.: The ovary of the guppy *Poecilia reticulata:* the atretic follicle, a corpus atreticum or a corpus luteum preovulations. Z. Zellforsch. *107:* 54–67 (1970b).

Lambert, J.G.D.: Steroidogenesis in the ovary of *Brachydanio rerio* (Teleostei); in Gaillard, Boer, Comparative endocrinology, pp. 65–68 (Elsevier/North-Holland Biomedical Press, Amsterdam 1978).

Lambert, J.G.D.; Mattheij, J.A.M.; Van Oordt, P.G.W.J.: The ovary and hypophysis of the zebra fish *Brachydanio rerio* during the reproductive cycle (Abstract). Gen. compar. Endocr. *18:* 602 (1972).

Lambert, J.G.D.; Van Oordt, P.G.W.J.: Ovarian hormones in teleosts. Fortschr. Zool. *22:* 340–349 (1974).

Lance, V.A.: MS thesis, College of William and Mary, Williamsburg, Va. (1968).

Lance, V.A.; Callard, K.P.: A histochemical study of ovarian function in the ovoviviparous elasmobranch, *Squalus acanthias.* Gen. compar. Endocr. *13:* 255–267 (1969).

Lange, R.H.: Highly conserved lipoprotein assembly in the teleost and amphibian yolk-platelet crystal. Nature, Lond. *289:* 329–330 (1981).

Lange, R.H.; Grodzinski, Z.; Kilarski, W.: Yolk-platelet crystals in 3 ancient bony fishes. *Polypterus bichir* (Polypteri), *Amia calva* and *Lepisosteus asseus* (Holostei). Cell tiss. Res. *222:* 159–166 (1982).

Larsen, L.O.: Subtotal hepatectomy in intact or hypophysectomized river lampreys *(Lampetra fluviatilis* L.): effects on regeneration, blood glucose regulation and vitellogenesis. Gen. compar. Endocr. *35:* 197–204 (1978).

Leger, C.; Fremont, L.; Marion, D.; Nassour, I.; Desfanges, M.-F.: Essential fatty acids in trout serum lipoproteins, vitellogenin and egg lipids. Lipids *16:* 593–600 (1981).

Lehri, G.K.: Cyclical changes in the ovary of catfish *Clarias batrachus* (Linn.). Acta anat. *69:* 105–124 (1968).

Le Menn, F.: Some aspects of vitellogenesis in a teleostean fish *Gobius niger.* L. Compar. Biochem. Physiol. *62A:* 495–499 (1979a).

Le Menn, F.: Induction de vitellogénine par l'oestradiol et par des androgènes chez un téléostéen marin: *Gobius niger* L. C.r. hebd. Séanc. Acad. Sci., Paris, Ser. D. *289:* 413–416 (1979b).

Le Menn, F.; Lamy, M.M.: Protéins vitellines sériques et hépatiques chez un poisson téléostéen, *Gobius niger* L., soumis à des traitements hormonaux. Bordeaux méd. *10:* 213–216 (1977).

Lessman, C.A.: Increase in acid phosphatase activity during fertilization of a teleost *Catostomus commersoni* egg. Experientia *37:* 415–416 (1981).

Liley, N.R.; Stacey, N.E.: Hormones, pheromones, and reproductive behaviour in fish; in Hoar, Randall, Donaldson, Fish physiology, vol. IX: Reproduction, part B, pp. 1–64 (Academic Press, New York 1983).

Livni, B.N.; Abraham, M.; Leray, C.; Yashouv, A.: Histochemical localization of hydroxysteroid dehydrogenases in the ovary of an euryhaline teleost fish, *Mugil capito.* Gen. compar. Endocr. *13:* 493 (1969).

Livni, N.: Ovarian histochemistry of the fishes *Cyprinus carpio, Mugil capito* and *Tilapia aurea* (Teleostei). Histochem. J. *3:* 405–414 (1971).

Lizenko, E.I.; Sidorov, V.S.; Nefedova, Z.A.; Potapova, O.I.: Content of lipid comporents in the eggs and milt of the cisco *Goregonus albula* (in Russian). Vopr. Ikhtiol. *19:* 368–370 (1979).

Lönning, S.: Comparative electron microscopic studies of teleostean eggs with special reference to the chorion. Sarasia *49:* 41–48 (1972).

Lönning, S.; Hagstrom, B.E.: Scanning electron microscope studies of the surface of the fish egg. Asarte *8:* 17–22 (1975).
Lönning, S.; Solemdal, P.: The relation between thickness of chorion and specific gravity of eggs from Norwegian and Baltic flat fish populations. FiskDir. Skr. Ser. Havunders. *16:* 77–87 (1972).
Lopes, R.A.; Leme Dos Santos, H.S.; Costa, J.R.V.; Pelizaro, M.G.; Costagnoli, N.: Histochemical study of oocyte zona radiata of the lambari *Astyanax bimaculatus lacustris* (Osteichthyes: Characidae). Zool. Anz. *208:* 265–268 (1982).
MacGregor, H.C.: The nucleolus and its genes in amphibian oogenesis. Biol. Rev. *47:* 177–210 (1972).
Makeeva, A.P.; Mikodina, E.V.: Structure of egg membranes in Cyprinidae and some data on their chemical nature. Biol. Nauki, Moscow *20:* 60–64 (1977).
Malhotra, Y.R.: On the nucleolar extrusions in the developing oocytes of a Kashmir fish *Schizothorax niger* (Heckel). Ichthyologica *11:* 57–60 (1963).
Malone, T.E.; Hisaoka, K.K.: A histochemical study of the formation of deutoplasmic components in developing oocytes of the zebra fish, *Brachydanio rerio.* J. Morph. *112:* 61–75 (1963).
Mani, K.; Saxena, P.K.: Effect of safe concentrations of some pesticides on ovarian recrudescence in the freshwater murrel, *Channa punctatus* (Bl.): a quantitative study. Ecotoxicol. envir. Saf. *92:* 241–249 (1985).
Manner, H.W.; Vancura, M.; Muehleman, C.: The ultrastructure of the chorion of the fathead minnow, *Pimephales promelas.* Trans. Am. Fish. Soc. *106:* 110–114 (1977).
Mano, Y.; Lipmann, F.: Characteristics of phosphoprotein (phosvitin) from a variety of fish roes. J. biol. Chem. *241:* 3822–3832 (1966).
Mano, Y.; Yoshida, M.: A novel composition of phosvitins from salmon and trout roe. J. Biochem. *66:* 105–108 (1969).
Markert, J.R.; Vanstone, W.E.: Egg proteins of the Pacific salmon, genus *Oncorhynchus:* proteins in the soluble fraction. Can. J. Biochem. *46:* 1334–1337 (1968).
Markert, J.R.; Vanstone, W.E.: Egg proteins of coho salmon *(Oncorhynchus kisutch)*: chromatographic separation and molecular weights of the major proteins in the high density fraction and their presence in salmon plasma. J. Fish. Res. Bd Can. *28:* 1853–1856 (1971).
Markov, K.P.: Stickiness of egg membranes in Acipenseridae (in Russian). Vopr. Ikhtiol. *18:* 483–493 (1978).
Masui, Y.; Clarke, H.J.: Oocyte maturation. Int. Rev. Cytol. *57:* 185–282 (1979).
Mazabraud, A.; Wegnez, M.; Denis, H.: Biochemical research on oogenesis. RNA accumulation in the oocytes of teleosts. Devl Biol. *44:* 326–332 (1975).
Mester, L.; Scripcariu, D.: Histochemical study of alkaline phosphatase and acid distribution in the oocytes of *Silurus glanis, Perca fluviatilis* and *Acerina cernua.* 1. Trans. Mus. Hist. Nat. Grigore Antipa *20:* 341–348 (1979).
Mester, R.; Scripcariu, D.; Varo, M.I.: Cytochemical localization of adenylate cyclase in fish ovary. Revue roum. Biol. Ser. Biol. Anim. *25:* 117–120 (1980).
Mikhailov, V.S.; Kostomarova, A.A.; Gulyanov, B.D.; Knyazeva, E.F.: Compartmentalization of DNA polymerases, α and γ in maturing oocytes of the loach, *Misgurnus fossilis.* Cell Differ. *13:* 87–92 (1983).
Miki, W.; Yamaguchi, K.; Konosu, S.: Comparison of carotenoids in the ovaries of marine fish and shellfish. Compar. Biochem. Physiol. *71B:* 7–12 (1982).

Mikodina, E.V.; Makeeva, A.P.: Structure and some characteristics of egg membranes in pelagophilous freshwater fishes. Vopr. Ikhtiol. *20:* 298–306 (1980).
Miller, O.L.; Beatty, B.R.: Visualization of nucleolar genes. Science *164:* 955–957 (1969).
Mirre, C.; Stahl, A.: Ulstrastructural study of nucleolar organizers in the quail oocyte during meiotic prophase. I. J. Ultrastruct. Res. *56:* 186–201 (1976).
Monaco, P.J.; Rasch, E.M.; Balsano, J.S.: Cytochemical and ultrastructural studies of oogenesis in a unisexual fish (Abstract). J. Histochem. Cytochem. *28:* 606 (1980).
Monaco, P.J.; Rasch, E.M.; Balsano, J.S.: Nucleoprotein cytochemistry during oogenesis in a unisexual fish, *Poecilia formosa.* Histochem. J. *13:* 747–762 (1981).
Müller, H.; Sterba, G.: Electronenmikroskopische Untersuchungen über Bildung und Struktur der Eihüllen bei Knochenfischen. II. Die Eihüllen jüngerer und älterer Oozyten von *Cynolebia belotti* Steindachner (Cyprinodontidae). Zool. Jb. Anat. *80:* 469–488 (1963).
Nagahama, Y.: The functional morphology of teleost gonads; in Hoar, Randa, Donaldson, Fish physiology, vol. IX: Reproduction, part A, pp. 223–276 (Academic Press, New York 1983).
Nagahama, Y.: Mechanism of gonadotropic control of steroidogenesis in the teleost ovarian follicle; in Gunma Symposia on Endocrinology, vol. 21, pp. 167–182 (Center for Academic Publications, Tokyo 1984).
Nagahama, Y.; Chan, K.C.; Hoar, W.S.: Histochemistry and ultrastructure of pre- and post-ovulatory follicles in the ovary of the gold fish, *Carassius auratus.* Can. J. Zool. *54:* 1128–1139 (1976).
Nagahama, Y.; Clarke, W.C.; Hoar, W.S.: Ultrastructure of putative steroid-producing cells in the gonads of coho *(Oncorhynchus kisutch)* and pink salmon *(Oncorhynchus gorbuscha).* Can. J. Zool. *56:* 2508–2519 (1978).
Nagahama, Y.; Hirose, K.; Young, G.; Adachi, S.; Suzuki, K.; Tamaoki, B.I.: Relative in vitro effectiveness of 17α-20β-dihydroxy-4-pregnen-3-one and other pregnene derivatives on germinal vesicle breakdown in oocytes of ayu *(Plecoglossus altivelis),* amago salmon *(Oncorhynchus rhodurus),* rainbow trout *(Salmo gairdneri),* and goldfish *(Carassius auratus).* Gen. compar. Endocr. *51:* 15–23 (1983b).
Nagahama, Y.; Kagawa, H.; Young, G.: Cellular sources of sex steroids in teleost gonads. Can. J. Fish. aquat. Sci. *39:* 56–64 (1982).
Nagahama, Y.; Young, G.; Kagawa, H.: Steroidogenesis in the amago salmon *(Oncorhynchus rhodurus)* ovarian follicle: a two cell-type model. Proc. 9th Int. Symp. Comp. Endocrinol., Hong Kong 1983a.
Nakano, E.: Fishes; in Metz, Monroy, Fertilization, vol. II, pp. 295–324 (Academic Press, New York 1969).
Narain, D.: Cytoplasmic inclusions in the oogenesis *Saccobranchus fossilis, Clarias batrachus* and *Anabas scandens.* Z. Zellforsch. *26:* 625 (1930).
Nath, V.; Bharpur, S.; Baker, A.: Fish oogenesis with particular reference to the so-called nucleolar extrusions. Proc. natn. Inst. Sci. India *20:* 247–253 (1944).
Nath, V.; Nangia, M.D.: A demonstration of the vacuole and the Golgi apparatus as independent cytoplasmic components in the fresh eggs of teleostean fishes. J. Morph. Physiol. *42:* 277–308 (1931).
Nath, V.; Sundararaj, B.I.: Isolation and identification of female specific serum lipophosphoprotein (vitellogenin) in the catfish *Heteropneustes fossilis* (Bloch). Gen. compar. Endocr. *43:* 184 (1981).

Nayyar, R.P.: The yolk nucleus of fish oocytes. Jl microsc. Sci. *105:* 351–358 (1964).
Neaves, W.B.: Leydig cells. Contraception *2:* 571–504 (1975).
Ng, T.B.; Campnell, C.M.; Idler, D.R.: Antibody inhibition of vitellogenesis and oocyte maturation in salmon and flounder. Gen. compar. Endocr. *41:* 233–239 (1980a).
Ng, T.B.; Idler, D.R.: Big and little forms of plaice vitellogenic and maturational hormones. Gen. compar. Endocr. *34:* 408–420 (1978).
Ng, T.B.; Idler, D.R.: Gonadotropic regulation of androgen production in flounder and salmonids. Gen. compar. Endocr. *42:* 25–38 (1980).
Ng, T.B.; Idler, D.R.: Yolk formation and differentiation in teleost fishes; in Hoar, Randall, Donaldson, Fish physiology, vol. IX: Reproduction, part B, pp. 373–404 (Academic Press, New York 1983).
Ng, T.B.; Idler, D.R.; Burton, M.P.: Effects of teleost gonadotropins and their antibodies on gonadal histology in winter flounder. Gen. compar. Endocr. *42:* 355–364 (1980b).
Nicholls, T.J.; Maple, G.: Ultrastructural observations on possible sites of steroid biosynthesis in the ovarian follicular epithelium of two species of cichlid fish, *Cichlasoma nigrofasciata* and *Haplochromis multicolor.* Z. Zellforsch. *128:* 317–335 (1972).
Norrevang, A.: Electron microscopic morphology of oogenesis. Int. Rev. Cytol. *23:* 113–185 (1968).
Nosek, J.: Biogenesis of the cortical granules in fish oocytes. Histochem. J. *16:* 435–437 (1984).
Nosek, J.; Krajhanzl, A.; Kocourek, J.: Binding of the cortical granule lectin to the jelly envelope in mature perch ova. Histochem. J. *16:* 429–431 (1984).
Ogata, H.; Nomura, T.; Hata, M.: Prostaglandin $F_{2\alpha}$ changes induced by ovulatory stimuli in the pond loach, *Misgurnus anguillicaudatus.* Bull. Jap. Soc. scient. fish. *45:* 929–931 (1979).
Ohta, H.; Teranishi, T.: Ultrastructure and histochemistry of granulosa and micropylar cells in the ovary of the loach, *Misgurnus anguillicaudatus* (Cantor). Bull. Fac. Fish. Hokkaido Univ. *33:* 1–8 (1982).
Osanai, K.: Scanning electron microscopy of the envelopes surrounding the chum-salmon oocytes. Bull. mar. Biol., St. Asamushi *16:* 21–25 (1977).
Oshiro, T.; Hibiya, T.: Presence of ovulation-inducing enzymes in the ovarian follicles of loach. Bull. Jap. Soc. scient. Fish. *41:* 115 (1975).
Oshiro, T.; Hibiya, T.: Water absorption of oocytes in the plaice, *Limanda yokohamae* during meiotic maturation and its role in rupturing follicles. Bull. Jap. Soc. scient. Fish. *47:* 835–841 (1981).
Oshiro, T.; Hibiya, T.: Protease secretion of ovarian follicles in the loach, *Misgurnus anguillicaudatus* suggesting the presence of ovulation-inducing enzymes. Bull. Jap. Soc. scient. Fish. *48:* 623–628 (1982).
Ozernyuk, N.D.: Activity of cytochrome oxidase in mitochondria of loach *(Misgurnus fossilis)* oocytes (in Russian). Izv. Akad. Nauk. SSSR Ser. Biol. *1981:* 136–138 (1981).
Ozernyuk, N.D.; Kotomin, A.V.: Changes in mitochondrial density during oogenesis of the loach (in Russian). Biokhimiya *41:* 2157–2160 (1976).
Ozernyuk, N.D.; Zviadadze, N.G.; Pal'Mbakh, L.R.: Respiration intensity, concentration and distribution of mitochondria in oogenesis of *Brachydanio rerio* (in Russian). Izv. Akad. Nauk. SSSR Ser. Biol. *1980:* 941–944 (1980).

Palombi, F.; Viron, A.: Nuclear cytochemistry of mouse oogenesis. 1. Changes in extra-nucleolar ribonucleoprotein components through meiotic prophase. J. Ultrastruct. Res. *61:* 10–20 (1977).

Pelizaro, M.G.; Leme Dos Santos, H.S.; Lopes, R.A.; Castagnoli, N.: Rhythm of development in the oocyte of the tambiu *Astyanax bimaculatus* (Pisces: Characidae). A morphometric and histochemical study. Archs Biol. *92:* 415–432 (1981).

Pendergrass, P.; Schroeder, P.: The ultrastructure of the thecal cell of the teleost, *Oryzias latipes,* during ovulation in vitro. J. Reprod. Fertil. *47:* 229–233 (1976).

Perry, J.M.: Post-fertilization changes in the chorion of winter flounder, *Pseudopleuronectes americanus* Walbaum, eggs observed with scanning electron microscope. J. Fish Biol. *25:* 83–94 (1984).

Peute, J.; Gaag, M.H. van der; Lambert, J.G.D.: Ultrastructure and lipid content of the liver of the zebra fish, *Brachydanio rerio,* related to vitellogenin synthesis. Cell Tiss. Res. *186:* 297–308 (1978).

Pickering, A.D.: Effects of gonadectomy, oestradiol and testosterone on the migrating river lamprey, *Lampetra fluviatilis* (L.). Gen. compar. Endocr. *28:* 473–480 (1976).

Plack, P.A.; Fraser, N.W.: Incorporation of *L*-^{14}C-leucine into egg proteins by liver slices from cod. Biochem. J. *121:* 857–862 (1971).

Plack, P.A.; Pritchard, D.J.; Fraser, N.W.: Egg proteins in cod serum: natural occurrence and induction by injections of oestradiol 3-benzoate. Biochem. J. *121:* 847–856 (1971).

Popov, D.; Gozia, O.; Caloianu-Iordachel, M.; Serban, M.: Comparative studies on proteins and enzymes from the oocyte of several teleostei species. Studii Cerc. Biochim. *20:* 71–76 (1977).

Rahn, R.H.; Sand, D.M.; Schlenk, H.: Wax esters in fish metabolism of dietary palmityl palmitate in the gourami *(Trichogaster cosby).* J. Nutr. *103:* 1441–1447 (1973).

Raikova, E.V.: Evolution of the nucleolar apparatus during oogenesis in Acipenseridae. J. Embryol. exp. Morph. *35:* 667–687 (1976).

Raikova, E.V.; Steinbert, G.; Thomas, C.: Localization of amplified ribosomal DNA in the meiotic prophase of acipenserid oocytes by in situ hybridization with ^{125}I-ribosomal RNA (in Russian). Tsitologiya *19:* 338–341 (1977).

Rajalakshmi, M.: Atresia of oocytes and ruptured follicles in *Gobius giuris* (Hamilton Buchanan). Gen. compar. Endocr. *6:* 378–384 (1966).

Ram, R.N.; Sathyanesan, A.G.: Effect of mercuric chloride on the reprodutive cycle of the teleostean fish *Channa punctatus.* Bull. env. Contam. Toxicol. *30:* 24–27 (1983).

Ramadan, A.A.; Ezzat, A.P.; Aziz, A.E.: Studies on fish oogenesis. 1. Histomorphological changes in the oocyte of *Merluccius merluccius* Mediterranean. Folia morph. *26:* 8–15 (1978).

Ramadan, A.; Ezzat, A.; Hafez, S.: Studies on fish oogenesis. 2. Cytological studies on developing oocytes of *Merluccius merluccius.* Folia morph. *27:* 18–24 (1979a).

Ramadan, A.; Ezzat, A.; Hafez, S.: Studies on fish oogenesis. 3. Cytochemical studies on developing oocytes of *Merluccius merluccius.* Folia morph. *27:* 182–186 (1979b).

Rasch, E.M.; Trehn, L.M.; Rasch, R.W.: Cytogenetic studies of *Poecilia pisces.* Chromosoma *31:* 18–40 (1970).

Rastogi, R.K.: Studies on fish oogenesis. 1. Histomorphological and cytochemical changes in the oocyte nucleus of *Amphipnous cuchia* (Ham.) Cytologia *33:* 357–369 (1968a).

Rastogi, R.K.: Occurrence and significance of follicular atresia in the catfish *Mystus tengara* (Ham.) Acta zool. *3:* 307–319 (1968b).
Rastogi, R.K.: The occurrence and significance of ovular atresia in the freshwater mud eel, *Amphipnous cuchia* (Ham.) Acta anat. *73:* 148–160 (1969a).
Rastogi, R.K.: Studies on the fish oogenesis. III. Vitellogenesis in some freshwater teleosts. Anat. Anz. *125:* 34–36 (1969b).
Rastogi, R.K.: Studies on the fish oogenesis. IV. Origin, structure and fate of the egg membranes in some freshwater teleosts. Acta biol. hung. *21:* 35–42 (1970).
Raven, C.P.: Oogenesis: the storage of developmental information (Pergamon Press, Oxford 1961).
Recher, L.; Whitescarver, J.; Briggs, L.: The fine structure of a nucleolar constituent. J. Ultastruct. Res. *29:* 1–14 (1969).
Reinboth, R.: Hormonal control of the teleost ovary. Zool. *12:* 307–324 (1972).
Richter, C.J.J.; Eding, F.H.; Leuven, S.E.W.; Wijst, J.G.M., van der: Effects of feeding levels and temperature on the development of the gonad in the African catfish *Clarias lazera* (C & V); in Richter, Goos, Reproductive physiology of fish. Proc. Int. Symp. on Reproductive Physiology of Fish, Wageningen 1982, pp. 147–149 (Centre for Agricultural Publishing and Documentation, Wageningen 1982).
Riehl, R.: Der Dotterkern in jungen Oocyten von *Noemacheilus barbatulus* (L.) (Teleostei, Cobitidae). Z. Naturforsch. *31C:* 361–362 (1976a).
Riehl, R.: Eine besondere Haftvorrichtung zwischen Cortex radiatus externus und Follikelepithel der Oocyten von *Gobio gobio* (L.) (Teleostei, Cyprinidae). Z. Naturforsch. *31C:* 628 (1976b).
Riehl, R.: Konzentrische 'Lamellen' in jungen Oocyten von *Noemacheilus barbatulus* (L.) (Teleostei, Cobitidae). Biol. Zbl. *86:* 523–528 (1977a).
Riehl, R.: Parakristalline Körper in jungen Oocyten der Schmerle *Noemacheilus barbatulus* (L.) (Teleostei, Cobitidae). Z. Naturforsch. *320:* 305–306 (1977b).
Riehl, R.: Histo- und ultrahistochemische Untersuchungen an den Oocyten von *Noemacheilus barbatulus* (L.) und *Gobio gobio* (L.) (Pisces, Teleostei). Zool. Anz., Jena *198:* 328–354 (1977c).
Riehl, R.: Der Follikel von *Noemacheilus barbatulus* (L.) und *Gobio gobio* (L.) und seine Beziehungen zur Oocyte. Zool. Anz., Jena *199:* 121–139 (1977d).
Riehl, R.: Licht- und elektronenmikroskopische Untersuchungen zu Bau und Entwicklung der Mikropylen von *Noemacheilus barbatulus* (L.) und *Gobio gobio* (L.) (Pisces, Teleostei). Zool. Anz., Jena *198:* 313–327 (1977e).
Riehl, R.: Elektronenmikroskopische und autoradiographische Untersuchungen an den Dotterkernen in den Oocyten von *Noemacheilus barbatulus* (L.) und *Phoxinus phoxinus* (L.) (Pisces, Teleostei). Cytobiologie *17:* 137–145 (1978a).
Riehl, R.: Die Oocyten der Grundel *Pomatoschistus minutus.* I. Licht- und elektronenmikroskopische Untersuchungen an Eihülle und Follikel. Helgoländer wiss. Meeresunters. *31:* 314–332 (1978b).
Riehl, R.: Licht- und elektronenmikroskopische Untersuchungen an den Oocyten der Süsswasser-Teleosteer *Noemacheilus barbatulus* (L.) und *Gobio gobio* (L.) (Pisces, Teleostei). Zool. Anz., Jena *201:* 199–219 (1978c).
Riehl, R.: Feinbau, Entwicklung und Bedeutung der Eihüllen bei Knochenfischen. Riv. ital. Piscic. Ittiop. *13A:* 113–121 (1978d).

Riehl, R.: Die Oocyten der Grundel *Pomatoschistus minutus.* II. Lichtmikroskopische Untersuchungen zur Kenntnis der Mikropyle. Microscopica Acta *80:* 287–291 (1978e).

Riehl, R.: Ein erweiterter und verbesserter Bestimmungschlüssel für die Eier deutscher Süsswasser-Teleosteer. Z. angew. Zool. *66:* 199–216 (1979).

Riehl, R.: Ultracytochemical localization of Na^+, K^+-activated ATPase in the oocytes of *Heterandria formosa Agassiz,* 1853 (Pisces, Poecillidae). Reprod. Nutr. Dev. *20:* 191–196 (1980a).

Riehl, R.: Micropyle of some salmonids and coregonids. Env. Biol. Fishes *5:* 59–66 (1980b).

Riehl, R.; Götting, K.J.: Zu Struktur und Vorkommen der Mikropyle an Eizellen und Eiern von Knochenfischen (Teleostei). Arch. Hydrobiol. *74:* 393–402 (1974).

Riehl, R.; Götting, K.J.: Bau und Entwicklung der Mikropylen in den Oocyten einiger Süsswasser-Teleosteer. Zool. Anz., Jena *195:* 363–373 (1975).

Riehl, R.; Schulte, E.: Licht- und elektronenmikroskopische Untersuchungen an den Eihüllen der Elritze (*Phoxinus phoxinus* (L.), Teleostei, Cyprinidae). Protoplasma *92:* 147–162 (1977a).

Riehl, R.; Schulte, E.: Vergleichende rasterelektronenmikroskopische Untersuchungen an den Mikropylen ausgewählter Süsswasser-Teleosteer). Arch. Fisch. Wiss. *28:* 95–107 (1977b).

Riehl, R.; Schulte, E.: Bestimmungsschlüssel der wichtigsten deutschen Süsswasser-Teleosteer anhand ihrer Eier. Arch. Hydrobiol. *83:* 200–212 (1978).

Roach, A.H.; Davies, P.L.: Catfish vitellogenin and its messenger RNA are smaller than their chicken and *Xenopus* counterparts. Biochim. biophys. Acta *610:* 400–412 (1980).

Robertson, D.A.: Possible functions of surface structure and size in some planktonic eggs of marine fishes. N. Z. J. mar. freshw. Res. *15:* 147–153 (1981).

Rogers, M.E.: Ribonucleoprotein particles in the amphibian oocyte nucleus. J. Cell Biol. *36:* 321–330 (1968).

Rubstov, V.F.: Membrane mucilage of carp (*Cyprinus carpio* L.) and herring (*Clupea harengus marisalbi* Berg) roe (in Russian). Vopr. Ikhtiol. *18:* 305–313 (1978).

Saidapur, S.K.: Follicular atresia in the ovaries of nonmammalian vertebrates. Int. Rev. Cytol. *5:* 225–244 (1978).

Saidapur, S.K.; Nadkarni, V.B.: Steroid synthesizing cellular sites in the ovary of catfish, *Mystus cavasius:* a histochemical study. Gen. compar. Endocr. *30:* 457–461 (1976).

Sakun, O.F.: Transition to meiotic division of oocytes with incomplete vitellogenesis under the influence of hormonal stimulation in the Atlantic salmon *Salmo salar* L. Dokl. Akad. Nauk SSSR *169:* 241–244 (1966).

Sakun, O.F.; Gureeva-Prebrazhenskaya, E.: A study of the maturation of the oocytes of teleostei in vitro, incubation methods and ways of analyzing the results. Vest. Leningr. Univ. Biol. *15:* 15–22 (1975).

Salmon, C.; Kagawa, H.; Adachi, S.; Nagahama, Y.; Fontaine, Y.-A.: Mise en évidence de liaison spécifique de la gonadotropine de Saumon chum *(Oncorhynchus keta)* dans des préparations membranaires de granulosa d'ovaire du Saumon amago *(Oncorhynchus rhodurus)* C.r. hebd. Séanc. Acad. Sci., Paris *298:* 337–340 (1984).

Sand, D.M.; Hehl, J.L.; Schlenk, H.: Biosynthesis of wax esters in fish: reduction of fatty acids and oxidation of alcohols. Biochemistry *8:* 4851–4854 (1969).

Sand, D.M.; Hehl, J.L.; Schlenk, H.: Biosynthesis of wax esters in fish. Metabolism of dietary alcohols. Biochemistry *10:* 2536–2541 (1971).

Sand, D.M.; Rahn, C.H.; Schlenk, H.: Wax esters in fish. Absorption and metabolism of oleyl alcohol in the gourami *(Trichogaster cosby).* J. Nutr. *103:* 600–607 (1973).

Saxena, P.K.; Anand, K.: A comparison of ovarian recrudescence in the catfish, *Mystus tengara* (Ham.) exposed to short photoperiods, to long photoperiods, and to melatonin. Gen. compar. Endocr. *33:* 506–511 (1977).

Saxena, O.P.; Bhatia, H.K.: Intranuclear localization of lipids in growing oocytes of a teleost, *Heteropneustes fossilis.* Ann. Zool., Agra *14:* 141–146 (1977).

Saxena, P.K.; Bhatia, R.: Effect of vegetable oil factory effluent on ovarian recrudescence in the freshwater teleost, *Channa punctatus* (Bl.). Water Air Soil Pollut. *20:* 55–61 (1983).

Scharrer, B.; Würzelmann, S.: Ultrastructural study on nuclear-cytoplasmic relationships in oocytes of the African lungfish *Protopterus aethiopicus.* 1. Nucleolo-cytoplasmic pathways. Z. Zellforsch. *96:* 325–343 (1969a).

Scharrer, B.; Würzelmann, S.: Ultrastructural study on nuclear-cytoplasmic relationships in oocytes of the African lungfish. *Protopterus aethiopicus.* II. The microtubular apparatus of the nuclear envelope. Z. Zellforsch. *101:* 1–12 (1969b).

Schmidt, G.; Bartsch, G.; Kitagawa, T.; Fujisawa, K.; Knolle, J.; Joseph, J.; De Marco, P.; Liss, M.; Haschemeyer, R.: Isolation of phosphoprotein of high phosphorus content from the eggs of brown brook-trout. Biochem. biophys. Res. Commun. *18:* 60–65 (1965).

Schreibman, M.P.; Berkowitz, E.J.; Hurk, R.V.D.: Histology and histochemistry of the testis and ovary of the platyfish, *Xiphophorus maculatus,* from birth to sexual maturity. Cell Tiss. Res *224:* 81–88 (1982).

Schroeder, P.C.; Pendergrass, P.: The inhibition of in vitro ovulation from follicles of the teleost, *Oryzias latipes,* by cytochalasin B. J. Reprod. Fertil. *48:* 327–330 (1976).

Schuetz, A.W.: Action of hormones on germinal vesicle breakdown in frog *(Rana pipiens)* oocytes. J. exp. Zool. *166:* 347–554 (1967).

Schulz, R.; Schlaghecke, R.; Bluem, V.: Variations of gonadal cyclic AMP content of male and female rainbow trout, *Salmo gairdneri* during the reproductive cycle. Compar. Biochem. Physiol. *70A:* 53–56 (1981).

Sehgal, A.; Sundararaj, B.I.: Effect of various photoperiodic regimens on the ovary of the cat-fish *Heteropneustes fossilis* (Bl.) during the spawning and the post-spawning periods. Biol. Reprod. *2:* 425–434 (1970).

Selman, K.; Wallace, R.A.: The inter- and intracellular passage of proteins through the ovarian follicle in teleosts; in Richter, Goos, Reproductive physiology of fish. Proc. Int. Symp. on Reproductive Physiology of Fish, Wageningen 1982, pp. 151–154 (Centre for Agricultural Publishing and Documentation, Wageningen 1982a).

Selman, K.; Wallace, R.A.: Oocyte growth in the sheepshead minnow *(Cyprinodon variegatus).* Uptake of exogenous proteins by vitellogenic oocytes. Tissue Cell *14:* 555–572 (1982b).

Semenov, V.V.: Dependence of vitellogenic oocytes growth rate in coregonid fishes upon functional condition of liver cells (in Russian). Arkh. anat. Gistol. Embriol. *78:* 98–107 (1980).

Seshachar, B.R.; Nayyar, R.P.: Intranuclear lipids in the early oocytes of *Heteropneustes fossilis* (Teleostei). Q. Jl microsc. Sci. *104:* 69–73 (1963).

Shackley, S.E.; King, P.E.: Oogenesis in a marine teleost, *Blennius pholis* L. Cell Tiss. Res. *181:* 105–128 (1977).

Shackley, S.E.; King, P.E.: Protein yolk synthesis in *Blennius pholis* L. J. Fish Biol. *13:* 179–193 (1978).

Shackley, S.E.; King, P.E.: Amino acid incorporation by isolated oocytes of the marine teleost *Blennius pholis* L. J. Fish Biol. *14:* 375–380 (1979).

Shaffi, S.A.; Jafri, A.K.: The activity of acid and alkaline phosphatase in the eggs of some freshwater teleosts. Indian J. Fish *21:* 296–298 (1975).

Shahi, N.R.P.; Mishra, A.P.; Singh, B.R.: Studies on the formation of compound yolk in the developing oocytes of an air-breathing fish, *Channa punctatus.* Cytologia, Tokyo *44:* 397–408 (1979).

Shanbag, A.B.; Nadkarni, V.B.: Identification of steroidogenic tissues in the ovary of a fish, *Channa gachua,* during different phases of reproductive cycle. J. Anim. Morphol. Physiol. *28:* 128–134 (1981).

Shelton, W.L.: Fate of the follicular epithelium in *Dorosoma petenense* (Pisces: Clupeidae). Copeia *1978:* 237–244 (1978).

Shrestha, T.K.: Rhythm of oocyte maturation and reproductive cycling in a female stone loach, *Neomacheilus beavani.* Int. J. Acad. Ichthyol *1:* 19–30 (1980).

Shukla, J.P.; Pandey, K.: Impaired ovarian functions in arsenic-treated freshwater fish, *Colisa fasciatus.* Toxicol. Lett., Amsterdam *20:* 1–4 (1984a).

Shukla, J.P.; Pandey, K.: Arsenic-induced structural changes during the ovarian cycle of a freshwater perch, *Colisa fasciatus.* Bull. Inst. Zool. Acad. Sin., Taipei *23:* 69–74 (1984b).

Simpson, T.H.; Wright, R.S.; Hunt, S.V.: Sex hormones in fish. II. The oestrogens of *Scyliorhinus caniculus.* J. Endocr. *26:* 499–507 (1963).

Singh, A.K.; Singh, T.P.: Effect of clomid, sexovid and prostaglandins on induction of ovulation and gonadotropin secretion in a freshwater catfish, *Heteropneustes fossilis* (Bloch). Endokrinologie *68:* 129–136 (1976).

Singh, B.N.; Boyle, W.: The vitellogenesis in *Gasterosteus aculeatus* (the stickleback) investigated by the ultracentrifuge. Q. Jl microsc. Sci. *81:* 83–106 (1938).

Singh, H.; Singh, T.P.: Effect of two pesticides on ovarian ^{32}P uptake and gonadotropin concentration during different phases of annual reproductive cycle in the freshwater catfish, *Heteropneustes fossilis* (Bloch). Environ. Res. *22:* 190–200 (1980a).

Singh, H.; Singh, T.P.: Effect of two pesticides on total lipid and cholesterol contents of the annual reproductive cycle in the freshwater teleost *Heteropneustes fossilis* (Bloch). Environ. Pollut. *23:* 9–17 (1980b).

Singh, H.; Singh, T.P.: Short-term effect of two pesticides on the survival, ovarian ^{32}P uptake and gonadotrophic potency in a freshwater catfish, *Heteropneustes fossilis* (Bloch). J. Endocr. *85:* 193–199 (1980c).

Singh, H.; Singh, T.P.: Effect of parathion and aldrin on survival, ovarian ^{32}P uptake and gonadotropic potency in freshwater catfish, *Heteropneustes fossilis* (Bloch). Endokrinologie *77:* 173–178 (1981).

Singh, H.; Singh, T.P.: Effect of two pesticides on hypothalamo-hypophysial-ovarian axis in the freshwater catfish, *Heteropneustes fossilis* (Bl.). Environ. Pollut. *27A:* 283–288 (1982).

Sobhana, B.; Nair, N.B.: Oogenesis in the olive carp, *Puntius sarana submasutus.* Zool. Anz. *198:* 373–379 (1977).

Sponaugle, D.L.: Annual fish oogenesis: morphology and pattern formation of the secondary egg envelope of *Cynolebias melanotaenis;* MS thesis, Clemson University (1980).

Srivastava, S.; Swarup, K.: Cyclic changes in the ovary of a freshwater large murrel *Channa striatus.* Archs Biol. *90:* 345–360 (1979).

Stacey, N.E.; Goetz, F.W.: Role of prostaglandins in fish reproduction. Can. J. Fish. aquat. Sci. *39:* 92–98 (1982).

Stahl, P.A.; Leray, C.: L'ovogenèse chez les poissons téléostéens. I. Origine et signification de la zona radiata et de ses annexes. Archs Anat. microsc. Morph. exp. *50:* 251–261 (1961).

Stehr, C.M.: The development of the hexagonally structured egg envelope of the C-O sole *(Pleuronichthys coenosus)*; MSc thesis, University of Washington (1982).

Stehr, C.M.; Hawkes, J.W.: The comparative ultrastructure of the egg membrane and associated pore structures in the starry flounder, *Platichthys stellatus* (Pallas), and pink salmon, *Oncorhynchus gorbuscha* (Walbaum). Cell Tiss. Res. *202:* 347–356 (1979).

Stehr, C.M.; Hawkes, J.W.: The development of the hexagonally structured egg envelope of the C-O sole *(Pleuronichthyes coenosus).* J. Morph. *178:* 267–284 (1983).

Stein, G.S.; Stein, J.L.: Recombinant DNA and cell proliferation (Academic Press, New York 1984).

Sterba, G.; Müller, H.: Elektronenmikroskopische Untersuchungen über Bildung und Struktur der Eihüllen bei Knochenfischen. I. Die Hüllen junger Oozyten von *Cynolebias belotti* steindachner (Cyprinodontidae). Zool. Jb. Anat. *80:* 65–80 (1962).

Stoss, J.: Fish gamete preservation and spermatozoan physiology; in Hoar, Randall, Donaldson, Fish physiology, vol. IX: Reproduction, part B, pp. 305–350 (Academic Press, New York 1983).

Subramaniam, M.K.; Aiyar, R.G.: Oogenesis of *Acentrogobius milli* with special reference to the behaviour of nucleoli. J. R. microsc. Soc. *55:* 174–183 (1935).

Sumida, B.Y.; Ahlstrom, E.H.; Moser, H.G.: Early development of seven flatfishes of the eastern north Pacific with heavily pigmented larvae (Pisces, Pleuronectiformes). Fishery Bull. *77:* 105–145 (1979).

Sun, J.; Xizai, W.: A preliminary study on the histological chemistry of the ovary of *Tilapia nilotica* during its development. Zool. Res. *4:* 279–286 (1983).

Sundararaj, B.I.; Goswami, S.V.: Effect of short and long term hypophysectomy on the ovary and inter-renal of cat-fish, *Heteropneustes fossilis* (Bl.). J. exp. Zool. *168:* 85–104 (1968).

Sundararaj, B.I.; Goswami, S.V.: Hormonal regulation of in vivo and in vitro oocyte maturation in the catfish, *Heteropneustes fossilis* (Bloch). Gen. compar. Endocr. *32:* 17–28 (1977).

Suyama, M.; Suzuki, T.: Free and conjugated amino acids in the extract of raibow trout egg. Bull. Jap. Soc. scient. Fish. *45:* 345–350 (1978).

Suzuki, T.; Suyama, M.: Properties of phosphopeptides in the extract of rainbow trout eggs. Bull. Jap. Soc. scient. Fish. *45:* 385–388 (1979).

Suzuki, K.; Tamaoki, B.-I.; Nagahama, Y.: In vitro synthesis of an inducer for germinal vesicle breakdown of fish oocytes, 17α,20β-dihydroxy-4-pregnen-3-one by ovarian tissue preparation of amago salmon *(Oncorhynchus rhodurus).* Gen. compar. Endocr. *45:* 533–535 (1981).

Szöllösi, D.; Billard, R.: The micropyle of trout eggs and its reaction to different incubation media. J. Microscopie *21:* 55–62 (1974).

Szöllösi, D.; Jalabert, B.: La thèque du follicule ovarien de la truite. J. Microsc., Paris *20:* 92a (1974).

Szöllösi, D.; Jalabert, B.; Breton, B.: Postovulatory changes in the theca folliculi of the trout. Ann. Biol. Anim. Biochim. Biophys. *18:* 883–892 (1978).

Takahashi, S.: The adherent filaments of eggs of isaza (*Chaenogobius isaza* Tanaka Pisces) Zool. Mag., Tokyo *87:* 216–220 (1978).

Takahashi, S.: Sexual maturity of the isoza, *Chaenogobius isaza.* 2. Gross morphology and histology of the ovary. Zool. Mag., Tokyo *90:* 54–61 (1981).

Takashima, Y.; Takano, K.; Hara, A.: Changes in female-specific proteins during the course of induced maturation of female Japanese eels *(Anguilla japonica).* Bull. Fac. Fish. Hokkaido Univ. *30:* 50–61 (1979).

Tang, F.; Lofts, B.; Chan, S.T.H.: Δ^5-3β-hydroxysteroid dehydrogenase activities in the ovary of the rice-field eel, *Monopterus albus.* Experientia *30:* 316–317 (1974).

Terkatin-Shimony, A.; Yaron, Z.: Estrogens and estrogenic effects in *Tilapia aurea* (Cichlidae, Teleostei). Ann. Biol. Anim. Biochim. Biophys. *18:* 1007–1012 (1978).

Tesoriero, J.V.: Formation of the chorion (zona pellucida) in the teleost, *Oryzias latipes.* I. Morphology of early oogenesis. J. Ultrastruct. Res. *59:* 282–291 (1977a).

Tesoriero, J.V.: Formation of the chorion (zona pellucida) in the teleost, *Oryzias latipes.* II. Polysaccharide cytochemistry of early oogenesis. J. Histochem. Cytochem. *25:* 1376–1380 (1977b).

Tesoriero, J.V.: Formation of the chorion (zona pellucida) in the teleost, *Oryzias latipes.* III. Autoradiography of (^{3}H)-proline incorporation. J. Ultrastruct. Res. *64:* 315–326 (1978).

Tesoriero, V.: The distribution and fate of tritium-labeled glucose and tritium-labeled galactose in oocytes of *Oryzias latipes:* formation of the cortical alveoli. Cell Tiss. Res. *209:* 117–130 (1980).

Theofan, G.; Goetz, F.W.: The in vitro effects of transcriptional and translational protein synthesis inhibitors on final maturation and ovulation of yellow perch *(Perca flavescens)* oocytes. Compar. Biochem. Physiol. *69A:* 557–561 (1981).

Thorgaard, G.H.: Chromosome set manipulation and sex control in fish; in Hoar, Randall, Donaldson, Fish physiology, vol. IX: Reproduction, part B, pp. 405–434 (Academic Press, New York 1983).

Thurston, R.V.: Electrophoretic patterns of blood serum proteins from rainbow trout *(Salmo gairdneri).* J. Fish. Res. Bd Can. *24:* 2169–2188 (1967).

Tokarz, R.R.: Oogonial proliferation, oogenesis, and folliculogenesis in nonmammalian vertebrates; in Jones, The vertebrate ovary, pp. 145–179 (Plenum Press, New York 1978).

Toshimori, K.; Oura, C.: An unusual architecture of occluding junctions between surface cells in teleost ovarian follicles *(Plecoglossus altivalis).* Archvm histol. jap. *42:* 543–550 (1979).

Toshimori, K.; Yasuzumi, F.: Tight junctions between ovarian follicle cells in the teleost *(Plecoglossus altivelis).* J. Ultrastruct. Res. *67:* 73–78 (1979a).

Toshimori, K.; Yasuzumi, F.: Gap junctions between microvilli of an oocyte and follicle cells in the teleost *(Plecoglossus altivelis).* Z. mikrosk.-anat. Forsch. *93:* 459–464 (1979b).

Toury, R.J.; Clérot, C.; André, J.: Les groupements mitochondriaux des cellules germinales des poissons téléostéens cyprinides. IV. Analyse biochimique des constituents du 'ciment' intermitochondrial isolé. Biol. cell. *30:* 225–231 (1977).

Toury, R.; Clérot, J.C.; Jean, A.: Mitochondrial clusters in cyprinid teleost fish *(Tinca tinca)* germ cells. Biochimie *61:* 405–414 (1979).

Treasurer, J.W.; Holiday, F.G.T.: Some aspects of the reproductive biology of perch *Perca fluviatilis:* a histological description of the reproductive cycle. J. Fish Biol. *18:* 359–376 (1981).

Tsukahara, J.: Ultrastructural study on the attaching filaments and villi of the oocyte of *Oryzias latipes* during oogenesis. Develop. Growth Different. *13:* 173–180 (1971).

Ueda, H.; Hiroi, O.; Hara, A.; Yamauchi, K.; Nagahama, Y.: Changes in serum concentrations of steroid hormones, thyroxine, and vitellogenin during spawning migration of the chum salmon, *Oncorhynchus keta.* Gen. compar. Endocr. *53:* 203–211 (1984).

Ullerich, F.H.: DNS-Gehalt und Chromosomenstruktur bei Amphibien. Chromosoma *30:* 1–37 (1970).

Ulrich, E.: Etude des ultrastructures au cours de l'ovogenèse d'un poisson téléostéen, le Danio, *Brachydanio rerio* (Hamilton Buchanan). J. Microscopie *8:* 447–478 (1969).

Upadhyay, S.N.: Morphologie des gonades immatures et étude expérimentale de l'induction de la gamétogenèse chez la truite arc-en-ciel juvenile (*Salmo gairdneri* R.). Dect. Etat. Univ. Paris-VI (1977).

Upadhyay, S.N.; Breton, B.; Billard, R.: Ultrastructural studies on experimentally induced vitellogenesis in juvernile rainbow trout (*Salmo gairdneri* R.). Ann. Biol. Anim. Biochim. Biophys. *18:* 1019–1025 (1978).

Utter, F.M.; Ridgeway, G.J.: A serologically detected serum factor associated with maturity in the English sole, *Parophyrys vitelus* and Pacific halibut, *Hippoglossus stenolepis.* Fishery Bull. *6:* 47–48 (1967).

Van Bohemen, C.G.; Lambert, J.G.D.: Estrogen synthesis in relation to estrone, estradiol, and vitellogenin plasma levels during the reproductive cycle of the female rainbow trout, *Salmo gairdneri.* Gen. compar. Endocr. *45:* 105–114 (1981).

Van Gansen, P.; Schram, A.: Evolution of the nucleoli during oogenesis in *Xenopus laevis* studied by electron microscopy. J. Cell Sci. *10:* 339–367 (1972).

Van Ree, G.E.; Lok, D.; Bosman, G.: In vitro induction of nuclear breakdown in oocytes of the zebra fish *Brachydanio rerio* (Ham. Buch). Effects of the composition of the medium and of protein and steroid hormones. Proc. K. ned. Akad. Wet. Ser. C *80:* 353–371 (1977).

Varma, S.K.: A comparative histochemical study of alkaline phosphatase in the vertebrate ovary. Acta morph. neerl. scand. *7:* 281–291 (1970).

Varma, S.K.; Guraya, S.S.: The localization and functional significance of alkaline phosphatase in the vertebrate ovary. Experientia *24:* 398–399 (1968).

Varo, M.I.; Mester, R.; Scripagriu, D.; Pecingine, L.: The presence of lysosomes in the oogenesis of *Carassius auratus* Gibelio. Revue roum. Biol. *22:* 27–32 (1977).

Veith, W.J.: The chemical composition of the follicular fluid of the viviparous teleost *Clinus superciliosus.* Compar. Biochem. Physiol. *63A:* 37–41 (1979).

Vincent, W.S.: Structure and chemistry of nucleoli. Int. Rev. Cytol. *4:* 269–295 (1955).

Vincent, W.S.; Halvorson, H.D.; Chen, H.R.; Shin, D.: A comparison of gene amplification in uni- and multinucleolate oocytes. Expl Cell Res. *57:* 340–350 (1969).

Vlad, M.: Nucleolar DNA in oocytes of *Salmo irideus* (Gibbons). Cell Tiss. Res. *167:* 407–424 (1976).
Vlaming, V.L. de: Environmental and endocrine control of teleost reproduction; in Schrecked, Control of sex in fishes, pp. 13–83 (Extension Division, Virginia Polytechnic Institute, State University, Blacksburg, 1974).
Vlaming, V.L. de; Wiley, H.S.; Delahunty, G.; Wallace, R.A.: Goldfish *(Carassius auratus)* vitellogenin: induction, isolation properties and relationship to yolk proteins. Compar. Biochem. Physiol. *67B:* 613–623 (1980).
Voss, W., Jr.; Fryer, J.L.; Banowetz, G.M.: Isolation, purification and partial characterization of a lectin from chinook salmon ova. Archs Biochem. Biophys. *186:* 25–34 (1978).
Wallace, W.: Observations on ovarian ova and follicles in certain teleostean and elasmobranch fishes. Q. Jl microsc. Sci. *47:* 161–215 (1903).
Wallace, R.A.: Oocyte growth in nonmammalian vertebrates; in Jones, The vertebrate ovary, pp. 469–502 (Plenum Press, New York 1978).
Wallace, R.A.; Jared, D.W.: Protein incorporation by isolated oocytes. V. Specificity for vitellogenin incorporation. J. Cell Biol. *69:* 345–351 (1976).
Wallace, R.A.; Jared, D.W.; Eisen, A.Z.: A general method for the isolation and purification of phosvitin from vertebrate egg. Can. J. Biochem. *44:* 1647–1652 (1966).
Wallace, R.A.; Selman, K.: Oogenesis in *Fundulus heteroclitus.* II. The transition from vitellogenesis into maturation. Gen. compar. Endocr. *42:* 345–354 (1980).
Wallace, R.A.; Selman, K.: Cellular and dynamic aspects of oocyte growth in teleosts. Am. Zool. *21:* 325–343 (1981).
Wegmann, I.: Götting, K.J.: Untersuchungen zur Dotterbildung in den Oocyten von *Xiphophorus helleri* (Heckel, 1948) (Teleostei, Poeciliidae). Z. Zellforsch. *119:* 405–433 (1971).
Wegnez, M.; Denis, H.; Mazabraud, A.; Clérot, J.C.: Biochemistry research on oogenesis. Devl Biol. *62:* 69–110 (1978).
Wheeler, J.P.G.: The growth of the egg in the dab, *Pleuronectes limada.* Q. Jl microsc. Sci. *68:* 641–660 (1924).
Wiegand, M.D.: Vitellogenesis in fishes; in Richter, Goos, Reproductive physiology of fish. Proc. Int. Symp. on Reproductive Physiology of Fish, Wageningen 1982, pp. 136–146 (Centre for Agricultural Publishing and Documentation, Wageningen 1982).
Wiegand, M.D.; Idler, D.R.: Synthesis of lipids by the rainbow trout *(Salmo gairdneri)* ovary in vitro. Can. J. Zool. *60:* 2683–2693 (1982).
Wischnitzer, S.: The submicroscopic morphology of the interphase nucleus. Int. Rev. Cytol. *34:* 1–45 (1973).
Wootton, R.J.: Environmental factors in fish reproduction; in Richter, Goos, Reproductive physiology of fish, pp. 210–219 (Centre for Agricultural Publishing and Documentation, Wageningen 1982).
Wourms, J.P.: Annual fish oogenesis. I. Differentiation of the mature oocyte and formation of the primary envelope. Devl Biol. *50:* 338–354 (1976).
Wourms, J.P.; Sheldon, H.: Annual fish oogenesis. II. Formation of the secondary egg envelope. Devl Biol. *50:* 355–366 (1976).
Yamamoto, K.: Studies on the formation of fish egg. V. The chemical nature and origin of the yolk vesicle in the oocytes of the herring, *Clupea pallasii.* Annotnes zool. jap. *28:* 158–162 (1955a).

Yamamoto, K. Studies on the formation of fish eggs. III. Localization of polysaccharides in oocytes of *Liopsetta obscura.* J. Fac. Sci. Hokkaido Univ. Ser. VI *12:* 391–399 (1956a).
Yamamoto, K.: Studies on the formation of fish eggs. VI. The chemical nature and the origin of yolk vesicle in the oocyte of the smelt, *Hypomesus japonicus.* Annotnes zool. jap. *28:* 233–237 (1955b).
Yamamoto, K.: Studies on the formation of fish eggs. VII. The fate of the yolk vesicle in the oocyte of the herring, *Clupea pallasii* during vitellogenesis. Annotnes zool. jap. *29:* 91–96 (1956b).
Yamamoto, K.: Studies on the formation of the fish eggs. VIII. The fate of the yolk vesicle in the oocyte of the smelt, *Hypomesus japonicus* during vitellogenesis. Embryologia *3:* 131–138 (1956c).
Yamamoto, K.: Studies on the formation of fish eggs. IX. The fate of the yolk vesicle in the oocytes of the flounder, *Liopsetta obscura,* during vitellogenesis. Bull. Fac. Fish. Hokkaido Univ. *7:* 208–212 (1956d).
Yamamoto, K.: Studies on the formation of fish eggs. I. Annual cycle in the development of ovarian eggs in flounder, *Leopsetta obscura.* J. Fac. Sci. Hokkaido Univ. Ser. 61 Zool. *12:* 362–373 (1956e).
Yamamoto, K.; Onozato, H.: Steroid-producing cells in the ovary of the zebra fish, *Brachydanio rerio.* Annotnes zool. jap. *41:* 119–128 (1968).
Yamamoto, K.; Oota, I.: An electron microscope study of the formation of the yolk globule in the oocyte of zebra fish *Brachydanio rerio.* Bull. Fac. Fish. Hokkaido Univ. *17:* 165–174 (1967a).
Yamamoto, K.; Oota, I.: Fine structure of yolk globules in the oocyte of the zebra fish, *Brachydanio rerio.* Annotnes zool. jap. *40:* 20–27 (1967b).
Yamamoto, K.; Oota, I.; Takano, K.; Ishikawa, T.: Studies on the maturing process of the rainbow trout, *Salmo gairdneri irideus.* 1. Maturation of the ovary of a one-year old fish. Bull. Jap. Soc. scient. Fish. *31:* 123–132 (1965).
Yamamoto, K.; Yamazaki, F.: Hormonal control of ovulation and spermiation in goldfish. Gunma Symp. Endocr. *4:* 131–145 (1967).
Yamamoto, M.: Electron microscopy of fish development. II. Oocyte-follicle cell relationship and formation of chorion in *Oryzias latipes.* J. Fac. Sci. Tokyo Sect. IV Zool. *10:* 123–127 (1963).
Yamamoto, M.: Electron microscopy of fish development. III. Changes in the ultrastructure of the nucleus and cytoplasm of the oocyte during its development in *Oryzias latipes.* J. Fac. Sci. Tokyo Univ. Sect. II *10:* 335–346 (1964).
Yamamoto, T.: Physiology of fertilization in fish eggs. Int. Rev. Cytol. *12:* 361–405 (1961).
Yamamoto, T.: Mechanism of breakdown of cortical alveoli during fertilization in medaka, *Oryzias latipes.* Embryologia *7:* 228–251 (1962).
Yamauchi, K.; Kagawa, H.; Ban, M.; Kasahara, N.; Nagahama, Y.: Changes in plasma estradiol-17β and 17α,20β-dihydroxy-4-pregnen-3-one levels during final oocyte maturation of the masu salmon *Oncorhynchus masou.* Bull. Jap. Soc. scient. Fish *50:* 2137 (1984).
Yamauchi, K.; Yamamoto, K.: Experiments on artificial maturation and fertilization of the Japanese eel *(Anguilla japonica);* in Richter, Goos, Reproductive physiology of fish, pp. 185–189 (Centre for Agricultural Publishing and Documentation, Wageningen 1982).

Yaron, Z.: Observations on the granulosa cells of *Acanthobrama terrae-sanctae* and *Tilapia nilotica* (Teleostei). Gen. compar. Endocr. *17:* 247–252 (1971).

Yorke, M.A.; McMillan, D.B.: Structural aspects of ovulation in the lamprey *Petromyzon marinus.* Biol. Reprod. *22:* 897–912 (1980).

Young, G.; Kagawa, H.; Nagahama, Y.: Inhibitory effect of cyanoketone on salmon gonadotropin-induced estradiol-17β production by ovarian follicles of the amago salmon *(Oncorhynchus rhodurus)* in vitro. Gen. compar. Endocr. *47:* 357–360 (1982a).

Young, G.; Kagawa, H.; Nagahama, Y.: Secretion of aromatizable Δ^4-androgens by thecal layers during estradiol-17β production by ovarian follicles of amago salmon *(Oncorhynchus rhodurus)* in vitro. Biomed. Res. *3:* 659–667 (1982b).

Young, G.; Kagawa, H.; Nagahama, Y.: Oocyte maturation in the amago salmon *(Oncorhynchus rhodurus):* in vitro effects of salmon gonadotropin, steroids and cyanoketone (an inhibitor of 3β-hydroxy-Δ^5-steroid dehydrogenase). J. exp. Zool. *224:* 265–275 (1982c).

Young, G.; Kagawa, H.; Nagahama, Y.: Evidence for a decrease in aromatase activity in the ovarian granulosa cells of amago salmon *(Oncorhynchus rhodurus)* associated with final oocyte maturation. Biol. Reprod. *29:* 310–315 (1983a).

Young, G.; Oeda, H.; Nagahama, Y.: Estradiol-17β and 17α,20β-dihydroxy-4-pregnen-3-one production by isolated ovarian follicles of amago salmon *(Oncorhynchus rhodurus)* in response to mammalian pituitary and placental hormones and salmon gonadotropin. Gen. compar. Endocr. *52:* 329–335 (1983b).

Yu, J.Y.-L.; Dickhoff, W.W.; Inui, Y.; Gorbman, A.: Sexual patterns of protein metabolism in liver and plasma of hagfish, *Eptatretus stouti* with special reference to vitellogenesis. Compar. Biochem. Physiol. *65B:* 111–117 (1980).

Yu, J.Y.-L.; Dickhoff, W.W.; Swanson, P.; Gorbman, A.: Vitellogenesis and its hormonal regulation in the Pacific hagfish, *Eptatratus stouti* L. Gen. compar. Endocr. *43:* 492–502 (1981).

Zahnd, J.P.; Porte, A.: Signes morphologiques de transfer de matériel nucléaire dans le cytoplasme des ovocytes de certaines espèces de poisson. C.r. hebd. Séanc. Acad. Sci., Paris *262D:* 1977–1978 (1966).

Zohar, Y.; Breton, B.; Billard, R.: Short-term profiles of plasma gonadotropin levels in the female rainbow trout throughout the reproductive cycle (Abstract). Gen. compar. Endocr. *46:* 396 (1982a).

Zohar, Y.; Breton, B.; Fostier, A.: Gonadotropic function during the reproductive cycle of the female rainbow trout, *Salmo gairdneri* in relation to ovarian steroid secretion: in vivo and in vitro studies; in Richter, Goos, Reproductive physiology of fish (Centre for Agricultural Publishing and Documentation, Wageningen 1982b).

Subject Index